THE OREXIN/ HYPOCRETIN SYSTEM

THE OREXIN/ HYPOCRETIN SYSTEM

Functional Roles and Therapeutic Potential

Edited by

JOSHUA A. BURK

JIM R. FADEL

ELSEVIER

ACADEMIC PRESS

An imprint of Elsevier

Academic Press is an imprint of Elsevier
125 London Wall, London EC2Y 5AS, United Kingdom
525 B Street, Suite 1650, San Diego, CA 92101, United States
50 Hampshire Street, 5th Floor, Cambridge, MA 02139, United States
The Boulevard, Langford Lane, Kidlington, Oxford OX5 1GB, United Kingdom

Library of Congress Cataloging-in-Publication Data
A catalog record for this book is available from the Library of Congress

British Library Cataloguing-in-Publication Data
A catalogue record for this book is available from the British Library

ISBN 978-0-12-813751-2

For information on all Academic Press publications
visit our website at https://www.elsevier.com/books-and-journals

Publisher: Nikki Levy
Acquisition Editor: Joslyn T. Chaiprasert-Paguio
Editorial Project Manager: Samuel Young
Production Project Manager: Surya Narayanan Jayachandran
Cover Designer: Vicky Pearson Esser

Typeset by SPi Global, India

CONTENTS

CONTRIBUTORS

Fernando Berrendero
Laboratory of Neuropharmacology, Department of Experimental and Health Sciences, Pompeu Fabra University, Barcelona; Faculty of Experimental Sciences, Francisco de Vitoria University, Madrid, Spain

Emily M. Black
Department of Neurobiology and Anatomy, Drexel University College of Medicine, Philadelphia, PA, United States

Sarah A. Blumenthal
Department of Psychological Sciences, College of William and Mary, Williamsburg, VA, United States

Joshua A. Burk
Department of Psychological Sciences, College of William and Mary, Williamsburg, VA, United States

Luis de Lecea
Department of Psychiatry and Behavioral Sciences, Stanford University, Stanford, CA, United States

Rodrigo A. España
Department of Neurobiology and Anatomy, Drexel University College of Medicine, Philadelphia, PA, United States

Jim R. Fadel
Department of Pharmacology, Physiology & Neuroscience, University of South Carolina School of Medicine, Columbia, SC, United States

África Flores
Institute of Neurosciences, Autonomous University of Barcelona, Barcelona, Spain

Kimberly J. Jennings
Department of Psychiatry and Behavioral Sciences, Stanford University, Stanford, CA, United States

Catherine M. Kotz
Department of Integrative Biology and Physiology, University of Minnesota; GRECC, Minneapolis VA Health Care System, Minneapolis, MN, United States

Jyrki P. Kukkonen
Department of Physiology, Institute of Biomedicine, Biomedicum Helsinki, and Biochemistry and Cell Biology, Faculty of Medicine, Department of Veterinary Biosciences, Faculty of Veterinary Medicine, University of Helsinki, Helsinki, Finland

Eden B. Maness
Department of Psychological Sciences, College of William and Mary, Williamsburg, VA, United States

Rémi Martin-Fardon
Department of Neuroscience, The Scripps Research Institute, La Jolla, CA, United States

Alessandra Matzeu
Department of Neuroscience, The Scripps Research Institute, La Jolla, CA, United States

Claudio Perez-Leighton
Department of Physiology, Faculty of Biological Sciences, Pontifical Catholic University of Chile, Chile

Gorica D. Petrovich
Department of Psychology, Boston College, Chestnut Hill, MA, United States

Jessica K. Shaw
Department of Neurobiology and Anatomy, Drexel University College of Medicine, Philadelphia, PA, United States

Jennifer A. Teske
Department of Nutritional Sciences, University of Arizona, Tucson, AZ, United States

Yanan Zhang
Research Triangle Institute, Research Triangle Park, NC, United States

PREFACE

In 1998, two research groups independently reported the discovery of novel neuropeptides—termed "hypocretins" or "orexins"—expressed in a restricted part of the brain encompassing the lateral hypothalamus, perifornical area, and dorsomedial hypothalamus. In the ensuing two decades since this discovery, there has been steady and growing progress to understand the variety of functions involving this neuropeptide system. Indeed, a PubMed search in 2018 using the term "hypocretin OR orexin" yields nearly 5000 hits—an astonishing number that reflects both the importance of these neuropeptides and the diversity of their physiological functions. The goal of this book is to provide an update about this progress after over 20 years of research in this area. To achieve this goal, we are fortunate to include chapters from many of the leading scientists who have studied orexins/hypocretins.

De Lecea played a lead role in the discovery of the hypocretins. In the first chapter, he and Jennings provide additional background on the orexin/hypocretin system and provide a broad overview of some of the relevant neuroanatomy and major research directions that have been spurred by their discovery. Kukkonen provides a comprehensive overview of the orexin/hypocretin receptor pharmacology, including endogenous signaling pathways and the growing array of drugs that act on either of the described orexin/hypocretin receptors. Collectively, these chapters provide the foundation in understanding the neural architecture of the orexin/hypocretin system. The book then moves to chapters focusing on the functions mediated by the orexin/hypocretin system.

Chapters by Kotz and colleagues and a chapter by Petrovich describe the role of orexins/hypocretins in feeding behavior and energy expenditure. These chapters describe the critical role of orexins/hypocretins in mediating feeding behavior—consistent with the localization of these neurons in the lateral hypothalamus—but also place the orexin system within a larger context of contributing to metabolism rates and body weight.

The role of the orexin/hypocretin system in motivated behavior is not restricted to feeding. More recently, orexins/hypocretins are recognized to have an important role in drug abuse mediated in part by extensive interactions between these neuropeptides and neural pathways classically implicated in reinforcement and reward, such as the mesocorticolimbic dopamine system. This topic is addressed by chapters contributed by España, focusing on dopaminergic signaling and by Matzeu & Martin-Fardon, focusing on cocaine addiction.

Another more recently recognized role for orexins is aspects of cognition and learning. The present authors include a chapter about orexins/hypocretins in cognition and Berrendero contributes a chapter about orexin/hypocretins regarding fear learning. Finally, we include some future directions for research regarding orexins/hypocretins, especially the development of compounds that may be useful for treating clinical conditions. Other topics, such as sleep, are discussed in multiple chapters in this book, as the changes in arousal associated with the orexin system affect a number of functions, from energy expenditure to cognition. Together, we hope that this book provides a useful, broad overview of many of the research areas and exciting developments that have followed the original description of orexins/hypocretins along with some ideas to stimulate future research.

Joshua A. Burk
Jim R. Fadel

CHAPTER 1

Hypocretins (Orexins): Twenty Years of Dissecting Arousal Circuits

Kimberly J. Jennings, Luis de Lecea
Department of Psychiatry and Behavioral Sciences, Stanford University, Stanford, CA, United States

ABBREVIATIONS

5-HT	serotonin
BF	basal forebrain
Ca^{2+}	calcium
CSF	cerebrospinal fluid
DA	dopamine
DREADDs	designer receptors exclusively activated by designer drugs
DRN	dorsal raphe nucleus
GABA	γ amino butyric acid
GPCR	G-protein-coupled receptor
GRIN	gradient refractive index
HA	histamine
Hcrt	hypocretin
HcrtR	hypocretin receptor
HPA	hypothalamic-pituitary-adrenal
Hz	Hertz
I.C.V.	interocerebroventricular
LC	locus coeruleus
LDT	laterodorsal tegmental nucleus
LH	lateral hypothalamus
MPN	median preoptic nucleus
mRNA	messenger ribonucleic acid
NAc	nucleus accumbens
NE	norepinephrine
NREM	nonrapid eye movement
POA	preoptic area
PPT	pedunculopontine tegmental nucleus
REM	rapid eye movement
TMN	tuberomammillary nucleus
VLPO	ventrolateral preoptic area
VTA	ventral tegmental area

The Orexin/Hypocretin System
https://doi.org/10.1016/B978-0-12-813751-2.00001-2

THE HYPOCRETINS: A BRIEF INTRODUCTION

The hypocretins (Hcrt, also known as orexins) were discovered by two independent research groups using complementary approaches (de Lecea et al., 1998; Sakurai et al., 1998). De Lecea and colleagues were seeking to identify specific genes enriched in the hypothalamus that might play a role in the wide-ranging functions performed by this heterogeneous brain region (de Lecea et al., 1998; Gautvik et al., 1996). They identified an mRNA whose expression was restricted to the perifornical region of the lateral hypothalamus (LH), and which was predicted to encode the precursor for two secreted peptides (Hcrt1 and Hcrt2, also known as orexin-A and orexin-B). These cells were observed to project widely throughout the brain and one of these peptides stimulated cultured hypothalamic neurons, consistent with a role in neurotransmission. These neuropeptides were named "hypocretins" due to their selective expression in the *hypo*thalamus and sequence similarity to the in*cretin* family of peptides (de Lecea et al., 1998).

Sakurai and colleagues were seeking to identify the natural ligands for orphaned GPCRs. They identified two peptides that each activated two related orphaned GPCRs (hypocretin/orexin receptor 1 and 2, or HcrtR1 and HcrtR2). They further demonstrated that I.C.V. infusion of either peptide in the rat increased food intake whereas fasting increased expression of these peptides. Thus they named these peptides "orexins," after the Greek *orexis* for appetite (Sakurai et al., 1998). However, both "hypocretin" and "orexin" refer to the same neuropeptides, and we now know that the function of Hcrt extends far beyond appetite regulation.

Hcrt neurons number approximately 5000 in rodents and 20,000–50,000 in humans (Date et al., 1999; Elias et al., 1998; Nambu et al., 1999; Peyron et al., 1998; Thannickal et al., 2000). Although Hcrt neurons are restricted to a small subregion of the LH, they project widely throughout the brain. Outside of the hypothalamus, the most robust projections are seen in the locus coeruleus (LC), thalamic paraventricular nuclei, periaqueductal gray, septum, bed nucleus of the stria terminalis, raphe peribrachial pontine region, medullary reticular formation, and the nucleus of the solitary tract (Chen, Dun, Kwok, Dun, & Chang, 1999; Date et al., 1999; Peyron et al., 1998). Many of these target regions are associated with arousal, hinting early on that Hcrt may play a role in sleep/wake regulation.

Hcrt neurons were quickly identified to be glutamatergic (Rosin, Weston, Sevigny, Stornetta, & Guyenet, 2003; Torrealba, Yanagisawa, & Saper, 2003). Hcrt neurons also coexpress a number of other substances, including dynorphin, galanin, neurotensin, proenkephalin, netafin, cocaine and amphetamine related transcript (CART), neuronal activity related-pentraxin (NARP), and protein delta-like 1 homolog (DLK-1) (Chou et al., 2001; Furutani et al., 2013; Hakansson, de Lecea, Sutcliffe, Yanagisawa, & Meister, 1999; Meister, Perez-Manso, & Daraio, 2013; Mickelsen et al., 2017; Reti, Reddy, Worley, & Baraban, 2002). Dynorphin is observed packaged with Hcrt in dense core vesicles within the VTA (Muschamp et al., 2014), but the potential corelease of other substances with Hcrt is unclear. Hcrt neurons are nonoverlapping with other sleep-regulating lateral hypothalamic neuropeptidergic populations, including melanin-concentrating hormone (MCH) and neuropeptide VF-expressing neurons (Peyron et al., 1998; Yelin-Bekerman et al., 2015).

Hcrt SIGNALING REGULATES AROUSAL

Hcrt Signaling Is Disrupted in Narcolepsy-Cataplexy

Soon after Hcrt's initial discovery, it was reported that knocking out the Hcrt gene in mice caused fragmented sleep/wake and behavioral arrests resembling cataplexy—a behavioral phenotype closely resembling narcolepsy in humans (Chemelli et al., 1999; Mochizuki et al., 2004). Mice with Hcrt neuronal degeneration show similarly disrupted sleep/wake and cataplexy-like attacks (Hara et al., 2001). Dogs with a heritable form of narcolepsy were also identified to have a mutation in the gene encoding HcrtR2 (Lin et al., 1999). These data suggested that disrupted Hcrt signaling may underlie narcolepsy in humans. Indeed, patients with narcolepsy have reduced or absent Hcrt1 in CSF (Nishino, Ripley, Overeem, Lammers, & Mignot, 2000) and an 85%–95% reduction in Hcrt neurons (Peyron et al., 2000; Thannickal et al., 2000). This was a landmark discovery in associating a restricted and nonredundant peptidergic system with disease, and low CSF Hcrt1 is now a diagnostic criterion for narcolepsy-cataplexy.

Researchers have evaluated the potential for viral gene transfer to rescue Hcrt signaling and attenuate narcoleptic symptoms in animal models. Widespread overexpression of Hcrt with an ectopically expressed transgene prevented cataplexy-like arrests and improved REM sleep abnormalities in Hcrt neuron-ablated mice (Mieda et al., 2004a), although widespread ectopic overexpression disrupts sleep/wake patterning in Hcrt-intact mice (Willie et al., 2011). Targeted transfer of the Hcrt gene to the LH (Liu et al., 2008), zona incerta (Liu et al., 2011), dorsolateral pons (Blanco-Centurion, Liu, Konadhode, Pelluru, & Shiromani, 2013), mediobasal hypothalamus (Kantor et al., 2013), and amygdala (Liu, Blanco-Centurion, Konadhode, Luan, & Shiromani, 2016) also improve symptoms of sleep fragmentation and cataplexy in Hcrt-deficient mice. Although application of such gene therapies in clinical settings remains a distant goal, these results suggest that Hcrt gene transfer may be a viable avenue for treatment to be considered as the technology matures.

Hcrt Promotes Wakefulness

Pharmacological studies in animal models have established a wake-promoting function for Hcrt. Central infusions of Hcrt increase wakefulness and decrease REM and NREM sleep (Hagan et al., 1999; Piper, Upton, Smith, & Hunter, 2000; Yamanaka et al., 2002), including local infusions into the LC (Bourgin et al., 2000; Hagan et al., 1999), basal forebrain (Espana, Baldo, Kelley, & Berridge, 2001; Methippara, Alam, Szymusiak, & McGinty, 2000; Thakkar, Ramesh, Strecker, & McCarley, 2001), tuberomammillary nucleus (TMN) (Huang et al., 2001), and pons (Xi, Fung, Yamuy, Morales, & Chase, 2002; Xi, Morales, & Chase, 2001). On the other hand, administration of HcrtR antagonists decrease wakefulness and increase both REM and NREM sleep (Brisbare-Roch et al., 2007; Winrow et al., 2011). Patterns of Hcrt neural activity and Hcrt release are consistent with a role for Hcrt in promoting arousal. Hcrt neuronal activation, CSF Hcrt1 concentration, and extracellular Hcrt1 concentrations are elevated during the natural active phase of both nocturnal and diurnal species (Estabrooke et al., 2001; Fujiki et al., 2001; Kiyashchenko et al., 2002; Modirrousta, Mainville, & Jones, 2005; Yoshida et al., 2001; Zeitzer et al., 2003). A higher temporal resolution description of Hcrt activity emerged in 2005, when Lee et al. and Mileykovskiy et al. independently described that Hcrt neurons in rats are phasically active during active waking, less active during quiet waking, and silent during sleep (Lee, Hassani, & Jones, 2005; Mileykovskiy, Kiyashchenko, & Siegel, 2005). It was further reported that Hcrt neurons ramp up activity shortly before wake in mice (Takahashi, Lin, & Sakai, 2008). In humans, Hcrt release within the amygdala was found to be elevated around wake onset relative to sleep (Blouin et al., 2013).

The development of optogenetic tools enabled manipulation of Hcrt neurons with ms precision to better understand the role of phasic Hcrt neural activity in promoting wake (Tyree & de Lecea, 2017a). Acute optogenetic stimulation of Hcrt neurons at frequencies above 5 Hz evoked wake from both REM and NREM sleep (Adamantidis, Zhang, Aravanis, Deisseroth, &

de Lecea, 2007). Hcrt peptides are essential for the wake-promoting effects of optogenetic stimulation, as Hcrt-deficient animals fail to show increased wake probability (Adamantidis et al., 2007). Furthermore, high sleep pressure blocks the wake-promoting effects of Hcrt stimulation (Carter, Adamantidis, Ohtsu, Deisseroth, & de Lecea, 2009), suggesting that sleep homeostasis gates the wake-promoting effects of Hcrt neuronal activity. In complementary studies, optogenetic inhibition of Hcrt neurons during the inactive phase using a halorhodopsin construct increased NREM sleep (Tsunematsu et al., 2011), whereas inhibition with an archaerhodopsin construct during the active phase increased total sleep and decreased wake (Tsunematsu et al., 2013). Consistent with these findings, chemogenetic manipulation of Hcrt neuronal activity using DREADDs, which act on a min-hr timescale, bidirectionally modulated time spent in wake and sleep (Sasaki et al., 2011). Thus Hcrt neural activity promotes sleep→wake transitions.

Differential Contributions of HcrtR1 and HcrtR2

Hcrt acts through two GPCRs to exert neuroexcitatory effects. HcrtR1 (also known as orexin receptor 1) has a higher affinity for Hcrt1 over Hcrt2 by an order of magnitude, whereas HcrtR2 (also known as orexin receptor 2) displays equal affinity for Hcrt1 and Hcrt2 (Sakurai et al., 1998). Consistent with Hcrt projections, HcrtRs are expressed widely throughout the brain. However, HcrtR1 and HcrtR2 have distinct if overlapping distributions, with some wake-promoting target regions primarily expressing Hcrt1 (e.g., LC, cholinergic neurons of the LDT and PPT), some expressing HcrtR2 (e.g., TMN), and some expressing both (e.g., DRN) (Lu, Bagnol, Burke, Akil, & Watson, 2000; Marcus et al., 2001; Trivedi, Yu, MacNeil, Van der Ploeg, & Guan, 1998).

Genetic loss-of-function studies reveal differential contributions of HcrtR1 and HcrtR2 to the regulation of arousal. HcrtR1 null mice exhibit relatively normal sleep/wake cycles (Hondo et al., 2009; Sakurai, 2007). In contrast, HcrtR2-deficient mice and dogs display sleep fragmentation and cataplexy-like arrests (Lin et al., 1999; Mochizuki et al., 2011; Willie et al., 2003), indicating that HcrtR2 signaling plays a more critical role in the stabilization of arousal states than Hcrt1-mediated pathways. However, HcrtR2 knockout produces a significantly milder behavioral phenotype than Hcrt or dual HcrtR1/HcrtR2 knockout (Hondo et al., 2009; Sakurai, 2007), suggesting both HcrtRs play a part in optimal functioning. Consistent with these findings, the effects of I.C.V. Hcrt1 on wake and NREM sleep are attenuated to a greater degree in Hcrt2 null mice than HcrtR1 null mice. However, Hcrt-mediated suppression of REM sleep was equally attenuated in both HcrtR1 and Hcrt2 null mice, indicating that both receptors are involved in Hcrt regulation of REM sleep (Mieda et al., 2011).

Substantial efforts to develop pharmaceutical treatments for narcolepsy, insomnia, and other sleep disturbances have generated numerous compounds targeting the Hcrt system. Among these, HcrtR-specific antagonists have been particularly useful for dissecting the contributions of HcrtR1 and HcrtR2 to arousal regulation. Consistent with genetic knockout studies, antagonism of HcrtR1 produces little to no changes in sleep time or architecture (Bonaventure et al., 2015; Dugovic et al., 2009, 2014; Gozzi et al., 2011; Morairty et al., 2012). Alternatively, HcrtR2 antagonism increases total sleep and decreases latency to sleep, primarily through increases in NREM sleep (Betschart et al., 2013; Bonaventure et al., 2015; Dugovic et al., 2009, 2014; Etori, Saito, Tsujino, & Sakurai, 2014; Kuduk et al., 2015; Morairty et al., 2012). This is in contrast to dual HcrtR antagonists, which greatly promote REM sleep (Brisbare-Roch et al., 2007; Dugovic et al., 2014; Winrow et al., 2011, 2012). Considered together, these data suggest that HcrtR1-mediated regulation of REM is only revealed in the absence of HcrtR2 signaling. Accordingly, HcrtR1 antagonists increased REM sleep if coadministered with a HcrtR2 antagonist (Dugovic et al., 2014) or if administered to a HcrtR2 null mouse (Bonaventure et al., 2017). Thus both HcrtRs are necessary for Hcrt-mediated suppression of REM, but HcrtR2 is primarily responsible for Hcrt-mediated suppression of NREM.

Hcrt INTERACTIONS WITH AROUSAL CIRCUITRY

Where in the brain does Hcrt act to promote arousal? As described earlier, Hcrt neurons project widely and target many arousal-regulating regions. Several of these regions also send reciprocal connections to Hcrt neurons, potentially participating in feedback loops. Because the Hcrt neuronal population is relatively small, spatially restricted, and its behavioral output so robust, investigations of Hcrt interactions with target regions have proven especially fruitful for elaborating arousal circuitry (Fig. 1).

Hcrt Interactions With the LC Noradrenergic System

The LC is likely a key effector point for Hcrt's wake-promoting effects. The LC, and particularly NE expressing neurons (LC^{NE}) therein, is a key wake-

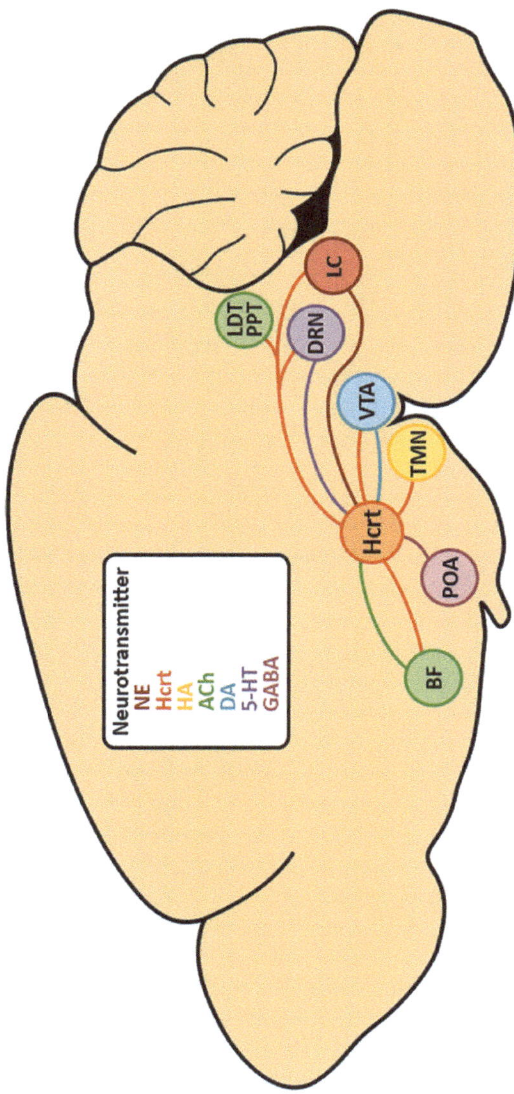

Fig. 1 Hcrt neurons project to and receive afferents from wake- and sleep-promoting centers of the brain.

promoting center displaying elevated firing during wake (Carter et al., 2010; Hobson, McCarley, & Wyzinski, 1975; Rasmussen, Morilak, & Jacobs, 1986). Hcrt neurons send intense projections to the LC (Peyron et al., 1998), where Hcrt acts to excite LC^{NE} neurons (Hagan et al., 1999; Horvath et al., 1999). Targeted infusions of Hcrt into the LC promote wakefulness (Bourgin et al., 2000; Espana et al., 2001; Hagan et al., 1999; Thakkar et al., 2001) and NE release into LC target regions (Walling, Nutt, Lalies, & Harley, 2004). Similarly, optogenetic stimulation of Hcrt neurons or Hcrt fibers within the LC activates LC neurons (Carter et al., 2009, 2012; Sears et al., 2013). Critically, inhibiting LC^{NE} neurons during Hcrt stimulation prevents Hcrt-evoked sleep-to-wake transitions in mice, and the wake-promoting effects of Hcrt overexpression or optogenetic stimulation are significantly attenuated in NE-deficient zebrafish (Carter et al., 2012; Singh, Oikonomou, & Prober, 2015). Conversely, depolarizing LC^{NE} neurons with a step-function opsin potentiates the effect of Hcrt stimulation, dramatically increasing the probability for sleep-to-wake transitions (Carter et al., 2012). Interestingly, Hcrt stimulation over 10–30 s increases wake probability whereas LC^{NE} stimulation exerts effects within 5 s. This is consistent with a role for the LC in mediating quick arousals when necessitated by environmental perturbations, whereas Hcrt may function as an upstream integrator of arousal-relevant cues over a slightly longer timescale (Carter, de Lecea, & Adamantidis, 2013).

Hcrt neuronal activity is modulated by NE input, suggesting a means for feedforward regulation of Hcrt activity. Adrenergic agonists induce hyperpolarization and decrease spike frequency in Hcrt neurons in vitro (Li, Gao, Sakurai, & van den Pol, 2002; Li & van den Pol, 2005; Yamanaka, Muraki, Tsujino, Goto, & Sakurai, 2003). Therefore LC^{NE}➔LH^{Hcrt} projections could exert negative feedback on Hcrt and downstream LC activity. Considering prolonged LC stimulation triggers behavioral arrests similar to cataplexy (Carter et al., 2010), negative feedback may be necessary for optimal functioning. NE also facilitates GABAergic neurotransmission onto Hcrt neurons in vitro, suggesting potential indirect Hcrt inhibition by LC^{NE} neurons (Li & van den Pol, 2005). However, there have also been reports of NE exciting Hcrt neurons (Bayer et al., 2005), and that sleep deprivation modulates the effects of NE on Hcrt electrophysiology in rats (Grivel et al., 2005; Uschakov et al., 2011). NE switching from stimulatory to inhibitory when the animal is under high sleep pressure may facilitate increased somnolence via suppressed Hcrt activity. However, such findings are limited at present to in vitro work and await confirmation in vivo. Importantly, the

LC does not send direct afferents to Hcrt neurons (Sakurai et al., 2005; Yoshida, McCormack, Espana, Crocker, & Scammell, 2006), indicating that LC^{NE} input in vivo is likely indirect or mediated by nonsynaptic mechanisms.

Hcrt Interactions With the VTA Dopaminergic System

The VTA has long been implicated in reward and reinforcement, but its role as an essential arousal-promoting region has only been recently demonstrated (Eban-Rothschild, Rothschild, Giardino, Jones, & de Lecea, 2016). The VTA is a major Hcrt projection target (Fadel & Deutch, 2002; Peyron et al., 1998). Although a relatively small proportion of Hcrt fibers form synaptic connections within the VTA, the presence of Hcrt-containing dense core vesicles along axons in the region suggests the potential for nonsynaptic release (Balcita-Pedicino & Sesack, 2007; Muschamp et al., 2014). Indeed, numerous studies have demonstrated that VTA dopaminergic neurons (VTA^{DA}) are activated by Hcrt in vitro and in vivo (Korotkova, Sergeeva, Eriksson, Haas, & Brown, 2003; Muschamp, Dominguez, Sato, Shen, & Hull, 2007; Nakamura et al., 2000; Narita et al., 2006; Uramura et al., 2001). Moreover, Hcrt facilitates glutamatergic and inhibits GABAergic neurotransmission onto VTA^{DA} neurons, influencing the excitatory:inhibitory balance of VTA^{DA} inputs (Baimel & Borgland, 2015; Borgland et al., 2009; Borgland, Storm, & Bonci, 2008; Borgland, Taha, Sarti, Fields, & Bonci, 2006; Tung et al., 2016). Hcrt also potentiates the VTA^{DA} response to several arousal-increasing cues, such as food (Cone, McCutcheon, & Roitman, 2014; Sheng, Santiago, Thomas, & Routh, 2014), cocaine (Espana, Melchior, Roberts, & Jones, 2011), and stress (Tung et al., 2016). Accordingly, intra-VTA Hcrt infusions increase DA release in VTA target regions (Narita et al., 2006; Vittoz & Berridge, 2006).

Hcrt is packaged and coreleased with dynorphin with variable effects on VTA^{DA} functioning. Disrupting Hcrt signaling results in an antireward phenotype which is rescued through concomitant disruption of dynorphin signaling, suggesting that the latent antireward properties of dynorphin signaling are normally masked by Hcrt (Muschamp et al., 2014). Consistent with this, Hcrt excites whereas dynorphin inhibits VTA^{DA} neurons (Muschamp et al., 2014). Additionally, a recent investigation revealed that Hcrt and dynorphin variably regulate VTA^{DA} activity depending on projection target. Hcrt selectively increased firing of VTA^{DA} neurons projecting to the NAc shell but not those projecting to the basolateral amygdala. On the other hand, dynorphin selectively decreased firing in VTA^{DA} neurons which projected to the medial NAc shell or the basolateral amygdala, but not the lateral NAc shell (Baimel, Lau, Qiao, & Borgland, 2017). Thus Hcrt/dynorphin release tunes VTA^{DA} activity in a projection-dependent manner. Whether such mechanisms interact with VTA-mediated arousal remain to be established.

Hcrt Interactions With the DRN Serotonergic System

Hcrt neurons also project to and excite wake-promoting 5-HT neurons of the DRN ($DRN^{5\text{-}HT}$) (Brown, Sergeeva, Eriksson, & Haas, 2001, 2002; Liu, van den Pol, & Aghajanian, 2002; Peyron et al., 1998; Takakusaki et al., 2005; Wang et al., 2005). Recently, it was reported that Hcrt application to $DRN^{5\text{-}HT}$ neurons in vitro increases high frequency input to these neurons by shifting the spectral power in membrane current noise, in addition to providing a slow depolarizing current as previously reported (Ishibashi et al., 2015; Takahashi et al., 2005). This Hcrt-mediated strengthening of high frequency input to $DRN^{5\text{-}HT}$ could support generation of the γ frequency (40–200 Hz) EEG oscillations observed during wake (Ishibashi et al., 2015). Hcrt was also reported to enhance the postspike afterhyperpolarization in $DRN^{5\text{-}HT}$ neurons, effectively increasing the high-pass filter properties of these neurons (Ishibashi et al., 2016). Thus Hcrt release during wake or arousing situations may enhance encoding of behaviorally relevant phasic inputs over tonic inputs to $DRN^{5\text{-}HT}$ neurons.

Hcrt-DRN interactions may also play a role in the generation of cataplexy. Targeted restoration of Hcrt2 into $DRN^{5\text{-}HT}$ neurons of dual HcrtR1/HcrtR2 null mice suppressed cataplexy-like arrests but had no effect on sleep/wake fragmentation, as did chemogenetic stimulation of $DRN^{5\text{-}HT}$ neurons (Hasegawa, Yanagisawa, Sakurai, & Mieda, 2014). Optogenetic stimulation of $DRN^{5\text{-}HT}$ terminals in the amygdala likewise suppressed cataplexy-like arrests in Hcrt-deficient mice, and optogenetic inhibition blocked the cataplexy-reducing effect of restoring HcrtR2 in $DRN^{5\text{-}HT}$ neurons (Hasegawa et al., 2017). These results indicate that a LH^{Hcrt}→$DRN^{5\text{-}HT}$→amygdala circuit regulates behavioral arrests resembling cataplexy.

Hcrt neurons likely receive additional negative feedback by 5-HT input. 5-HT application induces hyperpolarization in Hcrt neurons in vitro (Li & van den Pol, 2005; Muraki et al., 2004; Yamanaka, Muraki, et al., 2003). Hcrt neurons express several 5-HT receptor subtypes, although the 1a receptor subclass appears to mediate much of 5-HT's inhibitory effects (Jalewa et al., 2014; Muraki et al., 2004). DRN^{5HT} neurons may also indirectly regulate Hcrt activity, as optogenetic stimulation of 5-HT terminals in the LH inhibited Hcrt neurons both directly through the $5HT_{1a}$ receptor and indirectly through increased GABAergic transmission (Chowdhury & Yamanaka, 2016). 5-HT-mediated negative feedback mechanisms likely support maintenance of Hcrt neural activity within homeostatic boundaries.

Hcrt Interactions With the TMN Histaminergic System

Wake-promoting histaminergic neurons of the TMN (TMN^{HA}) are an important but likely not essential effector site for Hcrt-induced wakefulness. Hcrt excites TMN^{HA} neurons via HcrtR2 signaling (Bayer et al., 2005; Eriksson, Sergeeva, Brown, & Haas, 2001; Eriksson, Sergeeva, Selbach, & Haas, 2004; Willie et al., 2003; Yamanaka et al., 2002), whereas coreleased dynorphin supports TMN^{HA} excitation by suppressing GABAergic inputs (Eriksson et al., 2004). Intra-TMN Hcrt infusion promotes wakefulness and increases HA release in target regions (Huang et al., 2001), indicating that Hcrt can act on TMN^{HA} neurons to promote arousal. Furthermore, disruption of histamine-1-receptor signaling attenuates the wake-promoting effects of I.C.V. Hcrt administration (Huang et al., 2001; Shigemoto, Fujii, Shinomiya, & Kamei, 2004; Yamanaka et al., 2002). However, HA neurotransmission is not necessary for Hcrt neuron stimulation-mediated sleep-to-wake transitions. Optogenetic stimulation of Hcrt neurons successfully increased sleep-to-wake probability in HA-deficient mice and zebrafish (Carter et al., 2009; Chen, Singh, Oikonomou, & Prober, 2017). Thus although Hcrt actions on TMN^{HA} neurons are sufficient to promote wakefulness, their necessity for Hcrt-induced awakenings remains controversial and may depend on timescale of action.

Recent evidence indicates Hcrt neurons employ both Hcrt and glutamate to communicate separable information to TMN^{HA} neurons. Optogenetic stimulation of Hcrt terminals evoked two distinct patterns of TMN^{HA} firing in situ: an initial fast transient increase in spike frequency driven by glutamatergic input, and a HcrtR2-mediated slow-onset, long-lasting increase in firing which increased linearly over Hcrt stimulation (Schone, Apergis-Schoute, Sakurai, Adamantidis, & Burdakov, 2014). Thus coreleased glutamate and Hcrt may encode two distinct features of Hcrt neuron activity, with glutamate rapidly signaling changes in Hcrt neuron activity and Hcrt conveying the time integral of Hcrt neuronal input. Interestingly, similar computational components (derivative and integral controllers, respectively) can promote circuit stability when placed in a feedback loop, suggesting Hcrt/glutamate neurotransmission onto TMN^{HA} may support sleep/wake consolidation (Schone et al., 2014). However, such a feedback loop must include additional players as the TMN is not observed to send direct afferents to Hcrt neurons (Sakurai et al., 2005; Yoshida et al., 2006), nor does in vitro HA application alter Hcrt neuron activity (Tsujino et al., 2005; Yamanaka, Muraki, et al., 2003).

Hcrt Interactions With Cholinergic Systems

Cholinergic neurons of the BF, LDT, and PPT present additional arousal-regulating loci through which Hcrt can increase wakefulness. Hcrt neurons project to and excite cholinergic neurons within each of these regions (Burlet, Tyler, & Leonard, 2002; Chemelli et al., 1999; Eggermann et al., 2001; Peyron et al., 1998; Takahashi, Koyama, Kayama, & Yamamoto, 2002). Hcrt infusion into the BF or LDT increases wakefulness (Espana et al., 2001; Thakkar et al., 2001; Xi et al., 2001), indicating that Hcrt can act on the BF or LDT to promote arousal. Intra-PPT Hcrt inhibited REM atonia (Takakusaki et al., 2005), suggesting that disrupted Hcrt signaling in the PPT may contribute to the sudden and intrusive atonia of cataplexy. Hcrt also increases high frequency input onto LDT and PPT cholinergic neurons, as described previously for $DRN^{5\text{-}HT}$ (Ishibashi et al., 2015). Additionally, Hcrt induced a $Ca2+$-dependent high-frequency resonance in LDT/PPT cholinergic neurons extending up to 100 Hz (Ishibashi et al., 2015). These two Hcrt-mediated mechanisms may synergistically facilitate the encoding of high-frequency inputs onto LDT/PPT cholinergic neurons and therefore support high-frequency EEG oscillation during arousal and REM. Regarding reciprocal connections, cholinergic input to Hcrt neurons comes primarily from the BF, with few LDT/PPT neurons projecting to Hcrt neurons (Sakurai et al., 2005; Yoshida et al., 2006). In contrast to monoaminergic inputs, cholinergic input excites Hcrt neurons (Ohno, Hondo, & Sakurai, 2008; Yamanaka, Muraki, et al., 2003; Zhou, Gao, & Picciotto, 2015), suggesting a role in maintaining arousal states.

Hcrt Interactions With GABAergic Systems

Sleep-active GABAergic neurons within the POA, especially the VLPO and MPN, are thought to play a critical role in the initiation of NREM sleep and the maintenance of both NREM and REM sleep (Saper & Fuller, 2017; Schwartz & Kilduff, 2015). Hcrt neurons receive afferent input from these populations (Saito et al., 2013; Sakurai et al., 2005; Uschakov, Gong, McGinty, & Szymusiak, 2007; Yoshida et al., 2006), suggesting a means by which sleep-promoting regions can suppress arousal to facilitate wake→sleep transitions (Saper, Chou, & Scammell, 2001; Saper, Fuller, Pedersen, Lu, & Scammell, 2010). Consistent with this hypothesis, stimulation of POA^{GABA} neurons rapidly inhibited Hcrt neurons in situ and increased NREM sleep in vivo (Saito et al., 2013). Hcrt neurons also receive local GABAergic input, including from leptin-sensitive LH^{GABA} neurons (Bonnavion, Jackson, Carter, & de Lecea, 2015; Louis, Leinninger, Rhodes, & Myers Jr., 2010). GABAergic signaling to Hcrt neurons is essential for sleep/wake consolidation, as deleting $GABA_B$ receptors from Hcrt neurons causes severe sleep fragmentation without increasing total sleep or inducing cataplexy-like arrests (Matsuki et al., 2009).

Recently, it was demonstrated that inhibitory inputs (GABAergic or glycinergic) onto Hcrt neurons are modulated by sleep/wake history in zebrafish. Sleep deprivation reduced the number of inhibitory synapses onto Hcrt neurons during the active phase, but increased inhibitory synapses during inactive phase sleep rebound (Elbaz et al., 2017). These results complement an earlier study in mice demonstrating that sleep deprivation increased $GABA_A$ receptor expression and increased sensitivity to $GABA_A$ agonists in Hcrt neurons (Matsuki et al., 2015). Sleep deprivation also increased the total number of synapses onto Hcrt neurons in zebrafish (Appelbaum et al., 2010). Thus modulating structural plasticity within the LH is one mechanism by which homeostatic sleep pressure can regulate Hcrt neuron activity. Interestingly, whereas the total number of synapses onto Hcrt neurons exhibits a circadian rhythm (Appelbaum et al., 2010), fluctuations in the number of inhibitory synapses are circadian independent (Elbaz et al., 2017). Considered together, these data suggest that circadian mechanisms (e.g., clock gene transcriptional loops) may disproportionately modulate excitatory input onto Hcrt neurons, and further that circadian and homeostatic sleep pressure converge to modulate synaptic inputs onto Hcrt neurons.

Hcrt Interactions With Astrocytes

Increasingly, neuroscientists are coming to appreciate the functional contributions of astrocytes in regulating behavior, and sleep is no exception.

Optogenetic stimulation of astrocytes in the posterior hypothalamus decreased wake and increased both REM and NREM sleep (Pelluru, Konadhode, Bhat, & Shiromani, 2016), suggesting astrocytic regulation of arousal circuitry. Two recent studies have investigated how Hcrt neurons are modulated by astrocytes. Brain-wide or LH-restricted disruption of gap junctions within astrocytes fragmented wake and promoted sleep while also silencing Hcrt neurons. This was determined to be due to impaired glucose and lactate trafficking by astrocytes, highlighting Hcrt dependence on astrocytic support (Clasadonte, Scemes, Wang, Boison, & Haydon, 2017). In a separate study, sleep deprivation decreased expression of the astrocytic glutamate transporter 1 (GLT1) in close apposition with Hcrt neurons and increased GLT1 apposition with nearby sleep-promoting LH^{MCH} neurons. These changes were associated with increased tonic presynaptic inhibition of excitatory inputs onto Hcrt neurons and decreased postsynaptic response to high-frequency inputs in MCH neurons (Briggs, Hirasawa, & Semba, 2018). Thus astrocytes can differentially modulate the activity of Hcrt and MCH neurons within the LH in response to increased homeostatic sleep pressure.

Hcrt, BROADLY CONSIDERED

Hcrt: Role Within Sleep/Wake Circuitry

The essential nature of Hcrt for optimal sleep/wake and its hierarchical relationship with downstream effector regions makes Hcrt a logical starting point to develop models of sleep/wake regulation. Hcrt is particularly important for stabilizing sleep/wake transitions. This function is revealed most clearly through the narcolepsy-like fragmented sleep/wake phenotype of Hcrt-deficient animals (Chemelli et al., 1999; Lin et al., 1999). Indeed, increasing Hcrt neuronal activity precedes awakening (Lee et al., 2005; Mileykovskiy et al., 2005; Takahashi et al., 2008), and acute simulation evokes wake (Adamantidis et al., 2007; Carter et al., 2009, 2012). In the flip-flop model of sleep state switching, vigilance state is controlled by balanced activity between mutually inhibiting wake- and sleep-promoting loci (Saper et al., 2001, 2010; Saper, Scammell, & Lu, 2005). In this framework, Hcrt neurons support wake consolidation by increasing activity across multiple wake-promoting targets (e.g., LC^{NE}, VTA^{DA}, $DRN^{5\text{-}HT}$, TMN^{HA}, BF^{ACh}). Inhibitory input from sleep-promoting POA^{GABA} populations onto Hcrt neurons represents one pathway by which wakefulness can be suppressed to facilitate wake-to-sleep transitions. For more specific examinations of circuit interactions during sleep-to-wake transitions, computational frameworks have proven immensely useful for the generation and refinement of testable hypotheses (Carter et al., 2012). Indeed, quantitative models of sleep-to-wake transitions featuring Hcrt continue to provide explanatory and predictive value (Sorooshyari, Huerta, & de Lecea, 2015).

Hcrt in Arousal, Beyond Sleep and Wake

Evaluation of the role for Hcrt within arousal must also consider the influence of Hcrt on a multitude of other physiological systems including reward processing, stress reactivity, autonomic activity, and neuroendocrine function (Graebner, Iyer, & Carter, 2015; Li, Hu, & de Lecea, 2014). Briefly, Hcrt neurons are activated by a variety of stressful stimuli (Gonzalez, Iordanidou, Strom, Adamantidis, & Burdakov, 2016; Ida et al., 2000; Winsky-Sommerer et al., 2004), stimulate the HPA axis (Bonnavion et al., 2015; Hagan et al., 1999; Ida et al., 2000), and modulate anxiety-like behavior (Giardino & de Lecea, 2014; Heydendael, Sengupta, Beck, & Bhatnagar, 2014; Suzuki, Beuckmann, Shikata, Ogura, & Sawai, 2005), risk avoidance (Blomeley, Garau, & Burdakov, 2018), and impulsivity (Gentile et al., 2017; Muschamp et al., 2014). Hcrt also increases autonomic arousal with effects on heart rate, blood pressure, respiration, digestion, and body temperature (Balasko, Szelenyi, & Szekely, 1999; Carrive & Kuwaki, 2017; Samson, Gosnell, Chang, Resch, & Murphy, 1999; Shirasaka, Nakazato, Matsukura, Takasaki, & Kannan, 1999; Takahashi, Okumura, Yamada, & Kohgo, 1999). Disruptions to Hcrt signaling attenuate autonomic responses to stress (Kayaba et al., 2003; Kuwaki, 2011; Watanabe, Kuwaki, Yanagisawa, Fukuda, & Shimoyama, 2005; Zhang et al., 2010), suggesting that Hcrt is a key mediator for this component of stress reactivity. An expansive literature demonstrates that Hcrt also promotes reward-seeking behavior, including for drugs of abuse or natural rewards such as food or mating opportunities (Boutrel et al., 2005; Di Sebastiano & Coolen, 2012; James, Mahler, Moorman, & Aston-Jones, 2017; Tyree & de Lecea, 2017b). Furthermore, Hcrt regulates endocrine output within the gonadal, thyroid, prolactin, and growth hormone axes (Ferguson & Samson, 2003; Hagan et al., 1999; Inutsuka & Yamanaka, 2013). Critically, Hcrt neurons are highly sensitive to energetic homeostasis (Cai et al., 1999; Yamanaka et al., 2003). Hcrt activity is increased when animals are in need of energetic resources (Bonnavion et al., 2015; Gonzalez et al., 2016), potentially facilitating foraging and feeding behavior (Inutsuka et al., 2014; Mahler, Moorman, Smith, James, & Aston-Jones, 2014; Sakurai et al., 1998). Indeed, Hcrt neurons mediate fasting-induced increases in wakefulness (Yamanaka et al., 2003) and food-anticipatory increases in locomotor activity (Akiyama et al., 2004; Mieda et al., 2004b).

Hcrt as a Central Hub for Arousal

We conceptualize Hcrt as a critical integration node for the broad regulation of arousal (Fig. 2). Information regarding circadian and homeostatic sleep pressure, environmental stressors and rewards, and energetic status are all relevant to the decision to engage in arousal-dependent behaviors (e.g., goal-seeking, fight or flight, sleep). Hcrt neurons receive this information from multiple sources, each input having an associated weight. Input weights may physically manifest through relative expression of the multitude of neurotransmitter and neuropeptide receptors expressed on Hcrt neurons (Tsujino et al., 2005; Yamanaka et al., 2003; Yamanaka, Muraki, et al., 2003) or through changes in synaptic plasticity (Appelbaum et al., 2010; Elbaz et al., 2017), as described previously. Input weights can be considered zero sum, whereby increased weighting of one input class (e.g., sleep pressure) will decrease weighting of other(s) (e.g., "arousing" environmental stimuli) (Carter et al., 2009). An integrated representation of these data is output as phasic changes in Hcrt neuron firing. Integration of information by Hcrt neurons likely occurs over a multiple-seconds timescale, as stimulation of Hcrt neurons takes 10–30 s to increase sleep-to-wake probability (Adamantidis et al., 2007; Carter et al., 2013). Furthermore, different patterns of Hcrt neuronal activity may elicit differential release of Hcrt, glutamate, dynorphin, and likely other substances. These coreleased factors can modulate the impact of Hcrt signaling (Muschamp et al., 2014) or encode distinct components of Hcrt neuron information (Schone et al., 2014). Hcrt neuronal activity can thus facilitate optimal behavioral responding by promoting wake, autonomic arousal, and reward-seeking through widespread actions on diverse target regions. In keeping with this broad function, Hcrt neuron interactions with target regions can be modeled using control engineering theories as arousal control modules (Kosse & Burdakov, 2014; Schone & Burdakov, 2017).

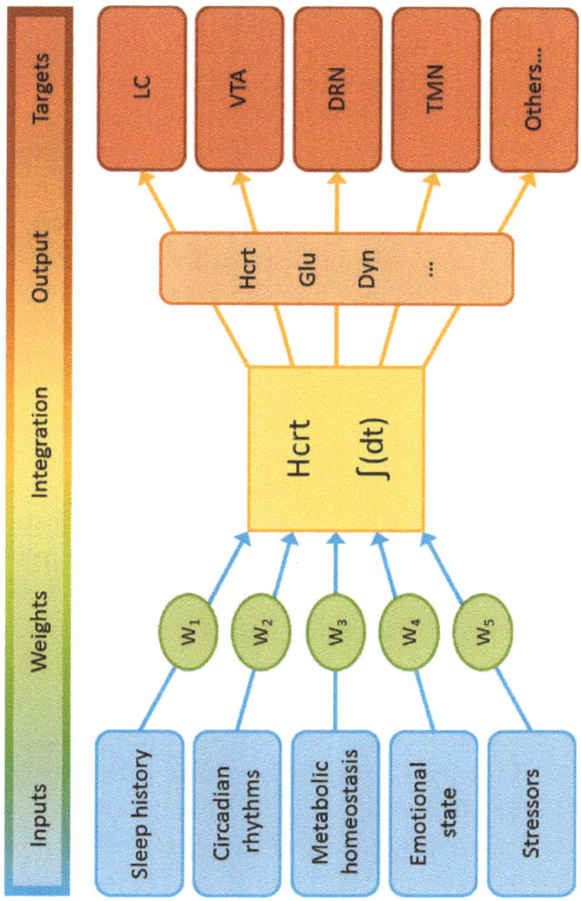

Fig. 2 Hcrt neurons act as an integrative hub for the control of arousal. Abbreviations: *W*, weight; *Glu*, glutamate; *Dyn*, dynorphin.

FUTURE DIRECTIONS

Therapeutic Advances

When Hcrt was identified to support wakefulness, it spurred a heroic effort to develop Hcrt-targeting pharmaceutical agents to treat sleep-related disturbances and disease. The first dual HcrtR antagonist achieved FDA approval for the treatment of primary insomnia in 2014, representing a remarkably quick trajectory for Hcrt from discovery to clinical target (Bennett, Bray, & Neville, 2014; Michelson et al., 2014). Additional small molecule HcrtR2 and dual HcrtR antagonists have reported positive outcomes in phase I and II clinical trials (Kumar, Chanana, & Choudhary, 2016). On the other hand, a nonpeptide HcrtR2 agonist was recently demonstrated to ameliorate narcolepsy-like symptoms in mice (Irukayama-Tomobe et al., 2017), inspiring hope that effective Hcrt therapy for narcolepsy-cataplexy is not long on the horizon. Recent description of the crystal structure of HcrtR2 will surely be of great assistance for the further development and refinement of Hcrt-targeting pharmaceuticals (Yin, Mobarec, Kolb, & Rosenbaum, 2015; Yin & Rosenbaum, 2017).

Heterogeneity Within the Hcrt System

A particularly interesting direction for future research is resolving the heterogeneity evident within the Hcrt population in terms of function and connectomics. Although some have suggested Hcrt neurons may be functionally segregated along the mediolateral axis (arousal regulating vs. reward regulating) (Harris & Aston-Jones, 2006), retrograde tracing reveals LC- and VTA-projecting Hcrt neurons are distributed across the Hcrt field (Gonzalez, Jensen, Fugger, & Burdakov, 2012). Yet, Hcrt neurons *can* be separated into at least two electrophysiologically distinct subgroups (Schone, Venner, Knowles, Karnani, & Burdakov, 2011; Williams, Alexopoulos, Jensen, Fugger, & Burdakov, 2008; Williams, Morton, & Burdakov, 2011). Recent efforts to characterize the transcriptome of Hcrt neurons with single-cell resolution have elegantly demonstrated the heterogeneous nature of this relatively small neuronal population (Yelin-Bekerman et al., 2015). It remains unclear whether such variation in genetic transcription indicates the existence of functional subpopulations. Nor is it clear whether subpopulations of Hcrt neurons are selectively targeted by specific afferents, or whether Hcrt subpopulations project with varying densities to target regions. Such circuit mapping could be highly informative regarding Hcrt's function in facilitating aspects of arousal/motivation under varying conditions. Application of recent advances in circuit mapping to the Hcrt system should prove illuminating (e.g., cTRIO (Beier et al., 2015; Schwarz et al., 2015)). Similarly, using genetically encoded Ca^{2+} reporters and recent advances in GRIN lenses to visualize Ca^{2+} activity within individual Hcrt neurons would provide invaluable information regarding Hcrt population activity and response dynamics.

Here's to the Next Twenty Years!

The last twenty years of Hcrt research have revealed more insights into sleep/wake circuitry than could ever have been predicted two decades ago. As technologies to manipulate and characterize neural circuits continue to mature, Hcrt will remain a particularly attractive foothold into arousal circuitry. By continuing to explore Hcrt's role in systems beyond sleep/wake, we move toward a more integrative understanding of hypothalamic function and of arousal itself. For all these reasons and more, we look forward to the next twenty years of Hcrt research!

ACKNOWLEDGMENTS

We thank Dr. Susan M. Tyree for her helpful comments during the preparation of this manuscript. This work was supported by grants from NIH (R01 MH087592, R01 MH102638) to LdL.

CONFLICTS OF INTEREST

The authors declare no conflicts of interest.

REFERENCES

Adamantidis, A. R., Zhang, F., Aravanis, A. M., Deisseroth, K., & de Lecea, L. (2007). Neural substrates of awakening probed with optogenetic control of hypocretin neurons. *Nature*, *450*, 420–424.

Akiyama, M., et al. (2004). Reduced food anticipatory activity in genetically orexin (hypocretin) neuron-ablated mice. *The European Journal of Neuroscience*, *20*, 3054–3062.

Appelbaum, L., et al. (2010). Circadian and homeostatic regulation of structural synaptic plasticity in hypocretin neurons. *Neuron*, *68*, 87–98.

Baimel, C., & Borgland, S. L. (2015). Orexin signaling in the VTA gates morphine-induced synaptic plasticity. *The Journal of Neuroscience*, *35*, 7295–7303.

Baimel, C., Lau, B. K., Qiao, M., & Borgland, S. L. (2017). Projection-target-defined effects of orexin and dynorphin on VTA dopamine neurons. *Cell Reports*, *18*, 1346–1355.

Balasko, M., Szelenyi, Z., & Szekely, M. (1999). Central thermoregulatory effects of neuropeptide Y and orexin A in rats. *Acta Physiologica Hungarica*, *86*, 219–222.

Balcita-Pedicino, J. J., & Sesack, S. R. (2007). Orexin axons in the rat ventral tegmental area synapse infrequently onto dopamine and gamma-aminobutyric acid neurons. *Journal of Comparative Neurology*, *503*, 668–684.

Bayer, L., et al. (2005). Opposite effects of noradrenaline and acetylcholine upon hypocretin/orexin versus melanin concentrating hormone neurons in rat hypothalamic slices. *Neuroscience*, *130*, 807–811.

Beier, K. T., et al. (2015). Circuit architecture of VTA dopamine neurons revealed by systematic input-output mapping. *Cell*, *162*, 622–634.

Bennett, T., Bray, D., & Neville, M. W. (2014). Suvorexant, a dual orexin receptor antagonist for the management of insomnia. *Pharmacy and Therapeutics*, *39*, 264–266.

Betschart, C., et al. (2013). Identification of a novel series of orexin receptor antagonists with a distinct effect on sleep architecture for the treatment of insomnia. *Journal of Medicinal Chemistry*, *56*, 7590–7607.

Blanco-Centurion, C., Liu, M., Konadhode, R., Pelluru, D., & Shiromani, P. J. (2013). Effects of orexin gene transfer in the dorsolateral pons in orexin knockout mice. *Sleep*, *36*, 31–40.

Blomeley, C., Garau, C., & Burdakov, D. (2018). Accumbal D2 cells orchestrate innate risk-avoidance according to orexin signals. *Nature Neuroscience*, *21*, 29–32.

Blouin, A. M., et al. (2013). Human hypocretin and melanin-concentrating hormone levels are linked to emotion and social interaction. *Nature Communications*, *4*, 1547.

Bonaventure, P., et al. (2015). A selective orexin-1 receptor antagonist attenuates stress-induced hyperarousal without hypnotic effects. *The Journal of Pharmacology and Experimental Therapeutics*, *352*, 590–601.

Bonaventure, P., et al. (2017). Evaluation of JNJ-54717793 a novel brain penetrant selective orexin 1 receptor antagonist in two rat models of panic attack provocation. *Frontiers in Pharmacology*, *8*, 357.

Bonnavion, P., Jackson, A. C., Carter, M. E., & de Lecea, L. (2015). Antagonistic interplay between hypocretin and leptin in the lateral hypothalamus regulates stress responses. *Nature Communications*, *6*, 6266.

Borgland, S. L., Storm, E., & Bonci, A. (2008). Orexin B/hypocretin 2 increases glutamatergic transmission to ventral tegmental area neurons. *The European Journal of Neuroscience*, *28*, 1545–1556.

Borgland, S. L., Taha, S. A., Sarti, F., Fields, H. L., & Bonci, A. (2006). Orexin A in the VTA is critical for the induction of synaptic plasticity and behavioral sensitization to cocaine. *Neuron*, *49*, 589–601.

Borgland, S. L., et al. (2009). Orexin A/hypocretin-1 selectively promotes motivation for positive reinforcers. *The Journal of Neuroscience*, *29*, 11215–11225.

Bourgin, P., et al. (2000). Hypocretin-1 modulates rapid eye movement sleep through activation of locus coeruleus neurons. *The Journal of Neuroscience*, *20*, 7760–7765.

Boutrel, B., et al. (2005). Role for hypocretin in mediating stress-induced reinstatement of cocaine-seeking behavior. *Proceedings of the National Academy of Sciences of the United States of America*, *102*, 19168–19173.

Briggs, C., Hirasawa, M., & Semba, K. (2018). Sleep deprivation distinctly alters glutamate transporter 1 apposition and excitatory transmission to orexin and MCH neurons. *The Journal of Neuroscience*.

Brisbare-Roch, C., et al. (2007). Promotion of sleep by targeting the orexin system in rats, dogs and humans. *Nature Medicine*, *13*, 150–155.

Brown, R. E., Sergeeva, O., Eriksson, K. S., & Haas, H. L. (2001). Orexin A excites serotonergic neurons in the dorsal raphe nucleus of the rat. *Neuropharmacology*, *40*, 457–459.

Brown, R. E., Sergeeva, O. A., Eriksson, K. S., & Haas, H. L. (2002). Convergent excitation of dorsal raphe serotonin neurons by multiple arousal systems (orexin/hypocretin, histamine and noradrenaline). *The Journal of Neuroscience*, *22*, 8850–8859.

Burlet, S., Tyler, C. J., & Leonard, C. S. (2002). Direct and indirect excitation of laterodorsal tegmental neurons by hypocretin/orexin peptides: Implications for wakefulness and narcolepsy. *The Journal of Neuroscience*, *22*, 2862–2872.
Cai, X. J., et al. (1999). Hypothalamic orexin expression: Modulation by blood glucose and feeding. *Diabetes*, *48*, 2132–2137.
Carrive, P., & Kuwaki, T. (2017). Orexin and central modulation of cardiovascular and respiratory function. *Current Topics in Behavioral Neurosciences*, *33*, 157–196.
Carter, M. E., Adamantidis, A., Ohtsu, H., Deisseroth, K., & de Lecea, L. (2009). Sleep homeostasis modulates hypocretin-mediated sleep-to-wake transitions. *The Journal of Neuroscience*, *29*, 10939–10949.
Carter, M. E., de Lecea, L., & Adamantidis, A. (2013). Functional wiring of hypocretin and LC-NE neurons: Implications for arousal. *Frontiers in Behavioral Neuroscience*, *7*, 43.
Carter, M. E., et al. (2010). Tuning arousal with optogenetic modulation of locus coeruleus neurons. *Nature Neuroscience*, *13*, 1526–1533.
Carter, M. E., et al. (2012). Mechanism for hypocretin-mediated sleep-to-wake transitions. *Proceedings of the National Academy of Sciences of the United States of America*, *109*, E2635–E2644.
Chemelli, R. M., et al. (1999). Narcolepsy in orexin knockout mice: Molecular genetics of sleep regulation. *Cell*, *98*, 437–451.
Chen, A., Singh, C., Oikonomou, G., & Prober, D. A. (2017). Genetic analysis of histamine signaling in larval zebrafish sleep. *eNeuro*, *4*.
Chen, C. T., Dun, S. L., Kwok, E. H., Dun, N. J., & Chang, J. K. (1999). Orexin A-like immunoreactivity in the rat brain. *Neuroscience Letters*, *260*, 161–164.
Chou, T. C., et al. (2001). Orexin (hypocretin) neurons contain dynorphin. *The Journal of Neuroscience*, *21*, RC168.
Chowdhury, S., & Yamanaka, A. (2016). Optogenetic activation of serotonergic terminals facilitates GABAergic inhibitory input to orexin/hypocretin neurons. *Scientific Reports*, *6*, 36039.
Clasadonte, J., Scemes, E., Wang, Z., Boison, D., & Haydon, P. G. (2017). Connexin 43-mediated astroglial metabolic networks contribute to the regulation of the sleep-wake cycle. *Neuron*, *95*, 1365–1380. e1365.
Cone, J. J., McCutcheon, J. E., & Roitman, M. F. (2014). Ghrelin acts as an interface between physiological state and phasic dopamine signaling. *The Journal of Neuroscience*, *34*, 4905–4913.
Date, Y., et al. (1999). Orexins, orexigenic hypothalamic peptides, interact with autonomic, neuroendocrine and neuroregulatory systems. *Proceedings of the National Academy of Sciences of the United States of America*, *96*, 748–753.
de Lecea, L., et al. (1998). The hypocretins: Hypothalamus-specific peptides with neuroexcitatory activity. *Proceedings of the National Academy of Sciences of the United States of America*, *95*, 322–327.
Di Sebastiano, A. R., & Coolen, L. M. (2012). Orexin and natural reward: Feeding, maternal, and male sexual behavior. *Progress in Brain Research*, *198*, 65–77.
Dugovic, C., et al. (2009). Blockade of orexin-1 receptors attenuates orexin-2 receptor antagonism-induced sleep promotion in the rat. *The Journal of Pharmacology and Experimental Therapeutics*, *330*, 142–151.
Dugovic, C., et al. (2014). Orexin-1 receptor blockade dysregulates REM sleep in the presence of orexin-2 receptor antagonism. *Frontiers in Neuroscience*, *8*, 28.
Eban-Rothschild, A., Rothschild, G., Giardino, W. J., Jones, J. R., & de Lecea, L. (2016). VTA dopaminergic neurons regulate ethologically relevant sleep-wake behaviors. *Nature Neuroscience*, *19*, 1356–1366.
Eggermann, E., et al. (2001). Orexins/hypocretins excite basal forebrain cholinergic neurones. *Neuroscience*, *108*, 177–181.

Elbaz, I., et al. (2017). Sleep-dependent structural synaptic plasticity of inhibitory synapses in the dendrites of hypocretin/orexin neurons. *Molecular Neurobiology, 54*, 6581–6597.
Elias, C. F., et al. (1998). Chemically defined projections linking the mediobasal hypothalamus and the lateral hypothalamic area. *Journal of Comparative Neurology, 402*, 442–459.
Eriksson, K. S., Sergeeva, O., Brown, R. E., & Haas, H. L. (2001). Orexin/hypocretin excites the histaminergic neurons of the tuberomammillary nucleus. *The Journal of Neuroscience, 21*, 9273–9279.
Eriksson, K. S., Sergeeva, O. A., Selbach, O., & Haas, H. L. (2004). Orexin (hypocretin)/dynorphin neurons control GABAergic inputs to tuberomammillary neurons. *The European Journal of Neuroscience, 19*, 1278–1284.
Espana, R. A., Baldo, B. A., Kelley, A. E., & Berridge, C. W. (2001). Wake-promoting and sleep-suppressing actions of hypocretin (orexin): Basal forebrain sites of action. *Neuroscience, 106*, 699–715.
Espana, R. A., Melchior, J. R., Roberts, D. C., & Jones, S. R. (2011). Hypocretin 1/orexin A in the ventral tegmental area enhances dopamine responses to cocaine and promotes cocaine self-administration. *Psychopharmacology, 214*, 415–426.
Estabrooke, I. V., et al. (2001). Fos expression in orexin neurons varies with behavioral state. *The Journal of Neuroscience, 21*, 1656–1662.
Etori, K., Saito, Y. C., Tsujino, N., & Sakurai, T. (2014). Effects of a newly developed potent orexin-2 receptor-selective antagonist, compound 1 m, on sleep/wakefulness states in mice. *Frontiers in Neuroscience, 8*, 8.
Fadel, J., & Deutch, A. Y. (2002). Anatomical substrates of orexin-dopamine interactions: Lateral hypothalamic projections to the ventral tegmental area. *Neuroscience, 111*, 379–387.
Ferguson, A. V., & Samson, W. K. (2003). The orexin/hypocretin system: A critical regulator of neuroendocrine and autonomic function. *Frontiers in Neuroendocrinology, 24*, 141–150.
Fujiki, N., et al. (2001). Changes in CSF hypocretin-1 (orexin A) levels in rats across 24 hours and in response to food deprivation. *Neuroreport, 12*, 993–997.
Furutani, N., et al. (2013). Neurotensin co-expressed in orexin-producing neurons in the lateral hypothalamus plays an important role in regulation of sleep/wakefulness states. *PLoS One, 8*, e62391.
Gautvik, K. M., et al. (1996). Overview of the most prevalent hypothalamus-specific mRNAs, as identified by directional tag PCR subtraction. *Proceedings of the National Academy of Sciences of the United States of America, 93*, 8733–8738.
Gentile, T. A., et al. (2017). Effects of suvorexant, a dual orexin/hypocretin receptor antagonist, on impulsive behavior associated with cocaine. *Neuropsychopharmacology*.
Giardino, W. J., & de Lecea, L. (2014). Hypocretin (orexin) neuromodulation of stress and reward pathways. *Current Opinion in Neurobiology, 29*, 103–108.
Gonzalez, J. A., Iordanidou, P., Strom, M., Adamantidis, A., & Burdakov, D. (2016). Awake dynamics and brain-wide direct inputs of hypothalamic MCH and orexin networks. *Nature Communications, 7*, 11395.
Gonzalez, J. A., Jensen, L. T., Fugger, L., & Burdakov, D. (2012). Convergent inputs from electrically and topographically distinct orexin cells to locus coeruleus and ventral tegmental area. *The European Journal of Neuroscience, 35*, 1426–1432.
Gonzalez, J. A., et al. (2016). Inhibitory interplay between orexin neurons and eating. *Current Biology, 26*, 2486–2491.
Gozzi, A., et al. (2011). Functional magnetic resonance imaging reveals different neural substrates for the effects of orexin-1 and orexin-2 receptor antagonists. *PLoS One, 6*, e16406.
Graebner, A. K., Iyer, M., & Carter, M. E. (2015). Understanding how discrete populations of hypothalamic neurons orchestrate complicated behavioral states. *Frontiers in Systems Neuroscience, 9*, 111.

Grivel, J., et al. (2005). The wake-promoting hypocretin/orexin neurons change their response to noradrenaline after sleep deprivation. *The Journal of Neuroscience, 25*, 4127–4130.
Hagan, J. J., et al. (1999). Orexin A activates locus coeruleus cell firing and increases arousal in the rat. *Proceedings of the National Academy of Sciences of the United States of America, 96*, 10911–10916.
Hakansson, M., de Lecea, L., Sutcliffe, J. G., Yanagisawa, M., & Meister, B. (1999). Leptin receptor- and STAT3-immunoreactivities in hypocretin/orexin neurones of the lateral hypothalamus. *Journal of Neuroendocrinology, 11*, 653–663.
Hara, J., et al. (2001). Genetic ablation of orexin neurons in mice results in narcolepsy, hypophagia, and obesity. *Neuron, 30*, 345–354.
Harris, G. C., & Aston-Jones, G. (2006). Arousal and reward: A dichotomy in orexin function. *Trends in Neurosciences, 29*, 571–577.
Hasegawa, E., Yanagisawa, M., Sakurai, T., & Mieda, M. (2014). Orexin neurons suppress narcolepsy via 2 distinct efferent pathways. *The Journal of Clinical Investigation, 124*, 604–616.
Hasegawa, E., et al. (2017). Serotonin neurons in the dorsal raphe mediate the anticataplectic action of orexin neurons by reducing amygdala activity. *Proceedings of the National Academy of Sciences of the United States of America, 114*, E3526–E3535.
Heydendael, W., Sengupta, A., Beck, S., & Bhatnagar, S. (2014). Optogenetic examination identifies a context-specific role for orexins/hypocretins in anxiety-related behavior. *Physiology & Behavior, 130*, 182–190.
Hobson, J. A., McCarley, R. W., & Wyzinski, P. W. (1975). Sleep cycle oscillation: Reciprocal discharge by two brainstem neuronal groups. *Science, 189*, 55–58.
Hondo, M., et al. (2009). Histamine-1 receptor is not required as a downstream effector of orexin-2 receptor in maintenance of basal sleep/wake states. *Acta Physiologica*, 287–294.
Horvath, T. L., et al. (1999). Hypocretin (orexin) activation and synaptic innervation of the locus coeruleus noradrenergic system. *Journal of Comparative Neurology, 415*, 145–159.
Huang, Z. L., et al. (2001). Arousal effect of orexin A depends on activation of the histaminergic system. *Proceedings of the National Academy of Sciences of the United States of America, 98*, 9965–9970.
Ida, T., et al. (2000). Possible involvement of orexin in the stress reaction in rats. *Biochemical and Biophysical Research Communications, 270*, 318–323.
Inutsuka, A., & Yamanaka, A. (2013). The physiological role of orexin/hypocretin neurons in the regulation of sleep/wakefulness and neuroendocrine functions. *Front Endocrinol (Lausanne), 4*(18).
Inutsuka, A., et al. (2014). Concurrent and robust regulation of feeding behaviors and metabolism by orexin neurons. *Neuropharmacology, 85*, 451–460.
Irukayama-Tomobe, Y., et al. (2017). Nonpeptide orexin type-2 receptor agonist ameliorates narcolepsy-cataplexy symptoms in mouse models. *Proceedings of the National Academy of Sciences of the United States of America, 114*, 5731–5736.
Ishibashi, M., et al. (2015). Orexin receptor activation generates gamma band input to cholinergic and serotonergic arousal system neurons and drives an intrinsic Ca(2+)-dependent resonance in LDT and PPT cholinergic neurons. *Frontiers in Neurology, 6*, 120.
Ishibashi, M., et al. (2016). Hypocretin/orexin peptides alter spike encoding by serotonergic dorsal raphe neurons through two distinct mechanisms that increase the late afterhyperpolarization. *The Journal of Neuroscience, 36*, 10097–10115.
Jalewa, J., et al. (2014). Neural circuit interactions between the dorsal raphe nucleus and the lateral hypothalamus: An experimental and computational study. *PLoS One, 9*, e88003.
James, M. H., Mahler, S. V., Moorman, D. E., & Aston-Jones, G. (2017). A decade of orexin/hypocretin and addiction: Where are we now? *Current Topics in Behavioral Neurosciences, 33*, 247–281.

Kantor, S., et al. (2013). Orexin gene therapy restores the timing and maintenance of wakefulness in narcoleptic mice. *Sleep*, *36*, 1129–1138.

Kayaba, Y., et al. (2003). Attenuated defense response and low basal blood pressure in orexin knockout mice. *American Journal of Physiology. Regulatory, Integrative and Comparative Physiology*, *285*, R581–R593.

Kiyashchenko, L. I., et al. (2002). Release of hypocretin (orexin) during waking and sleep states. *The Journal of Neuroscience*, *22*, 5282–5286.

Korotkova, T. M., Sergeeva, O. A., Eriksson, K. S., Haas, H. L., & Brown, R. E. (2003). Excitation of ventral tegmental area dopaminergic and nondopaminergic neurons by orexins/hypocretins. *The Journal of Neuroscience*, *23*, 7–11.

Kosse, C., & Burdakov, D. (2014). A unifying computational framework for stability and flexibility of arousal. *Frontiers in Systems Neuroscience*, *8*, 192.

Kuduk, S. D., et al. (2015). Identification of MK-8133: An orexin-2 selective receptor antagonist with favorable development properties. *Bioorganic & Medicinal Chemistry Letters*, *25*, 2488–2492.

Kumar, A., Chanana, P., & Choudhary, S. (2016). Emerging role of orexin antagonists in insomnia therapeutics: An update on SORAs and DORAs. *Pharmacological Reports*, *68*, 231–242.

Kuwaki, T. (2011). Orexin links emotional stress to autonomic functions. *Autonomic Neuroscience*, *161*, 20–27.

Lee, M. G., Hassani, O. K., & Jones, B. E. (2005). Discharge of identified orexin/hypocretin neurons across the sleep-waking cycle. *The Journal of Neuroscience*, *25*, 6716–6720.

Li, J., Hu, Z., & de Lecea, L. (2014). The hypocretins/orexins: Integrators of multiple physiological functions. *British Journal of Pharmacology*, *171*, 332–350.

Li, Y., Gao, X. B., Sakurai, T., & van den Pol, A. N. (2002). Hypocretin/Orexin excites hypocretin neurons via a local glutamate neuron: A potential mechanism for orchestrating the hypothalamic arousal system. *Neuron*, *36*, 1169–1181.

Li, Y., & van den Pol, A. N. (2005). Direct and indirect inhibition by catecholamines of hypocretin/orexin neurons. *The Journal of Neuroscience*, *25*, 173–183.

Lin, L., et al. (1999). The sleep disorder canine narcolepsy is caused by a mutation in the hypocretin (orexin) receptor 2 gene. *Cell*, *98*, 365–376.

Liu, M., Blanco-Centurion, C., Konadhode, R. R., Luan, L., & Shiromani, P. J. (2016). Orexin gene transfer into the amygdala suppresses both spontaneous and emotion-induced cataplexy in orexin-knockout mice. *The European Journal of Neuroscience*, *43*, 681–688.

Liu, M., et al. (2008). Orexin (hypocretin) gene transfer diminishes narcoleptic sleep behavior in mice. *The European Journal of Neuroscience*, *28*, 1382–1393.

Liu, M., et al. (2011). Orexin gene transfer into zona incerta neurons suppresses muscle paralysis in narcoleptic mice. *The Journal of Neuroscience*, *31*, 6028–6040.

Liu, R. J., van den Pol, A. N., & Aghajanian, G. K. (2002). Hypocretins (orexins) regulate serotonin neurons in the dorsal raphe nucleus by excitatory direct and inhibitory indirect actions. *The Journal of Neuroscience*, *22*, 9453–9464.

Louis, G. W., Leinninger, G. M., Rhodes, C. J., & Myers, M. G., Jr. (2010). Direct innervation and modulation of orexin neurons by lateral hypothalamic LepRb neurons. *The Journal of Neuroscience*, *30*, 11278–11287.

Lu, X. Y., Bagnol, D., Burke, S., Akil, H., & Watson, S. J. (2000). Differential distribution and regulation of OX1 and OX2 orexin/hypocretin receptor messenger RNA in the brain upon fasting. *Hormones and Behavior*, *37*, 335–344.

Mahler, S. V., Moorman, D. E., Smith, R. J., James, M. H., & Aston-Jones, G. (2014). Motivational activation: A unifying hypothesis of orexin/hypocretin function. *Nature Neuroscience*, *17*, 1298–1303.

Marcus, J. N., et al. (2001). Differential expression of orexin receptors 1 and 2 in the rat brain. *Journal of Comparative Neurology, 435*, 6–25.
Matsuki, T., et al. (2009). Selective loss of GABA(B) receptors in orexin-producing neurons results in disrupted sleep/wakefulness architecture. *Proceedings of the National Academy of Sciences of the United States of America, 106*, 4459–4464.
Matsuki, T., et al. (2015). GABAA receptor-mediated input change on orexin neurons following sleep deprivation in mice. *Neuroscience, 284*, 217–224.
Meister, B., Perez-Manso, M., & Daraio, T. (2013). Delta-like 1 homologue is a hypothalamus-enriched protein that is present in orexin-containing neurones of the lateral hypothalamic area. *Journal of Neuroendocrinology, 25*, 617–625.
Methippara, M. M., Alam, M. N., Szymusiak, R., & McGinty, D. (2000). Effects of lateral preoptic area application of orexin-A on sleep-wakefulness. *Neuroreport, 11*, 3423–3426.
Michelson, D., et al. (2014). Safety and efficacy of suvorexant during 1-year treatment of insomnia with subsequent abrupt treatment discontinuation: A phase 3 randomised, double-blind, placebo-controlled trial. *Lancet Neurology, 13*, 461–471.
Mickelsen, L. E., et al. (2017). Neurochemical heterogeneity among lateral hypothalamic hypocretin/orexin and melanin-concentrating hormone neurons identified through single-cell gene expression analysis. *eNeuro, 4*, .
Mieda, M., et al. (2004a). Orexin peptides prevent cataplexy and improve wakefulness in an orexin neuron-ablated model of narcolepsy in mice. *Proceedings of the National Academy of Sciences of the United States of America, 101*, 4649–4654.
Mieda, M., et al. (2004b). Orexin neurons function in an efferent pathway of a food-entrainable circadian oscillator in eliciting food-anticipatory activity and wakefulness. *The Journal of Neuroscience, 24*, 10493–10501.
Mieda, M., et al. (2011). Differential roles of orexin receptor-1 and -2 in the regulation of non-REM and REM sleep. *The Journal of Neuroscience, 31*, 6518–6526.
Mileykovskiy, B. Y., Kiyashchenko, L. I., & Siegel, J. M. (2005). Behavioral correlates of activity in identified hypocretin/orexin neurons. *Neuron, 46*, 787–798.
Mochizuki, T., et al. (2004). Behavioral state instability in orexin knock-out mice. *The Journal of Neuroscience, 24*, 6291–6300.
Mochizuki, T., et al. (2011). Orexin receptor 2 expression in the posterior hypothalamus rescues sleepiness in narcoleptic mice. *Proceedings of the National Academy of Sciences of the United States of America, 108*, 4471–4476.
Modirrousta, M., Mainville, L., & Jones, B. E. (2005). Orexin and MCH neurons express c-Fos differently after sleep deprivation vs. recovery and bear different adrenergic receptors. *The European Journal of Neuroscience, 21*, 2807–2816.
Morairty, S. R., et al. (2012). Dual hypocretin receptor antagonism is more effective for sleep promotion than antagonism of either receptor alone. *PLoS One, 7*, e39131.
Muraki, Y., et al. (2004). Serotonergic regulation of the orexin/hypocretin neurons through the 5-HT1A receptor. *The Journal of Neuroscience, 24*, 7159–7166.
Muschamp, J. W., Dominguez, J. M., Sato, S. M., Shen, R. Y., & Hull, E. M. (2007). A role for hypocretin (orexin) in male sexual behavior. *The Journal of Neuroscience, 27*, 2837–2845.
Muschamp, J. W., et al. (2014). Hypocretin (orexin) facilitates reward by attenuating the antireward effects of its cotransmitter dynorphin in ventral tegmental area. *Proceedings of the National Academy of Sciences of the United States of America, 111*, E1648–E1655.
Nakamura, T., et al. (2000). Orexin-induced hyperlocomotion and stereotypy are mediated by the dopaminergic system. *Brain Research, 873*, 181–187.
Nambu, T., et al. (1999). Distribution of orexin neurons in the adult rat brain. *Brain Research, 827*, 243–260.

Narita, M., et al. (2006). Direct involvement of orexinergic systems in the activation of the mesolimbic dopamine pathway and related behaviors induced by morphine. *The Journal of Neuroscience, 26*, 398–405.

Nishino, S., Ripley, B., Overeem, S., Lammers, G. J., & Mignot, E. (2000). Hypocretin (orexin) deficiency in human narcolepsy. *Lancet, 355*, 39–40.

Ohno, K., Hondo, M., & Sakurai, T. (2008). Cholinergic regulation of orexin/hypocretin neurons through M(3) muscarinic receptor in mice. *Journal of Pharmacological Sciences, 106*, 485–491.

Pelluru, D., Konadhode, R. R., Bhat, N. R., & Shiromani, P. J. (2016). Optogenetic stimulation of astrocytes in the posterior hypothalamus increases sleep at night in C57BL/6J mice. *The European Journal of Neuroscience, 43*, 1298–1306.

Peyron, C., et al. (2000). A mutation in a case of early onset narcolepsy and a generalized absence of hypocretin peptides in human narcoleptic brains. *Nature Medicine, 6*, 991–997.

Peyron, C., et al. (1998). Neurons containing hypocretin (orexin) project to multiple neuronal systems. *The Journal of Neuroscience, 18*, 9996–10015.

Piper, D. C., Upton, N., Smith, M. I., & Hunter, A. J. (2000). The novel brain neuropeptide, orexin-A, modulates the sleep-wake cycle of rats. *The European Journal of Neuroscience, 12*, 726–730.

Rasmussen, K., Morilak, D. A., & Jacobs, B. L. (1986). Single unit activity of locus coeruleus neurons in the freely moving cat I. During naturalistic behaviors and in response to simple and complex stimuli. *Brain Research, 371*, 324–334.

Reti, I. M., Reddy, R., Worley, P. F., & Baraban, J. M. (2002). Selective expression of Narp, a secreted neuronal pentraxin, in orexin neurons. *Journal of Neurochemistry, 82*, 1561–1565.

Rosin, D. L., Weston, M. C., Sevigny, C. P., Stornetta, R. L., & Guyenet, P. G. (2003). Hypothalamic orexin (hypocretin) neurons express vesicular glutamate transporters VGLUT1 or VGLUT2. *Journal of Comparative Neurology, 465*, 593–603.

Saito, Y. C., et al. (2013). GABAergic neurons in the preoptic area send direct inhibitory projections to orexin neurons. *Front Neural Circuits, 7*, 192.

Sakurai, T., et al. (1998). Orexins and orexin receptors: A family of hypothalamic neuropeptides and G protein-coupled receptors that regulate feeding behavior. *Cell, 92*, 573–585.

Sakurai, T., et al. (2005). Input of orexin/hypocretin neurons revealed by a genetically encoded tracer in mice. *Neuron, 46*, 297–308.

Sakurai, T. (2007). The neural circuit of orexin (hypocretin): Maintaining sleep and wakefulness. *Nature Reviews Neuroscience, 8*, 171–181.

Saper, C. B., Chou, T. C., & Scammell, T. E. (2001). The sleep switch: Hypothalamic control of sleep and wakefulness. *Trends in Neurosciences, 24*, 726–731.

Saper, C. B., & Fuller, P. M. (2017). Wake-sleep circuitry: An overview. *Current Opinion in Neurobiology, 44*, 186–192.

Saper, C. B., Fuller, P. M., Pedersen, N. P., Lu, J., & Scammell, T. E. (2010). Sleep state switching. *Neuron, 68*, 1023–1042.

Saper, C. B., Scammell, T. E., & Lu, J. (2005). Hypothalamic regulation of sleep and circadian rhythms. *Nature, 437*, 1257–1263.

Samson, W. K., Gosnell, B., Chang, J. K., Resch, Z. T., & Murphy, T. C. (1999). Cardiovascular regulatory actions of the hypocretins in brain. *Brain Research, 831*, 248–253.

Sasaki, K., et al. (2011). Pharmacogenetic modulation of orexin neurons alters sleep/wakefulness states in mice. *PLoS One, 6*, e20360.

Schone, C., Apergis-Schoute, J., Sakurai, T., Adamantidis, A., & Burdakov, D. (2014). Coreleased orexin and glutamate evoke nonredundant spike outputs and computations in histamine neurons. *Cell Reports, 7*, 697–704.

Schone, C., Venner, A., Knowles, D., Karnani, M. M., & Burdakov, D. (2011). Dichotomous cellular properties of mouse orexin/hypocretin neurons. *The Journal of Physiology, 589*, 2767–2779.

Schone, C., & Burdakov, D. (2017). Orexin/hypocretin and organizing principles for a diversity of wake-promoting neurons in the brain. *Current Topics in Behavioral Neurosciences*, *33*, 51–74.
Schwartz, M. D., & Kilduff, T. S. (2015). The neurobiology of sleep and wakefulness. *The Psychiatric Clinics of North America*, *38*, 615–644.
Schwarz, L. A., et al. (2015). Viral-genetic tracing of the input-output organization of a central noradrenaline circuit. *Nature*, *524*, 88–92.
Sears, R. M., et al. (2013). Orexin/hypocretin system modulates amygdala-dependent threat learning through the locus coeruleus. *Proceedings of the National Academy of Sciences of the United States of America*, *110*, 20260–20265.
Sheng, Z., Santiago, A. M., Thomas, M. P., & Routh, V. H. (2014). Metabolic regulation of lateral hypothalamic glucose-inhibited orexin neurons may influence midbrain reward neurocircuitry. *Molecular and Cellular Neurosciences*, *62*, 30–41.
Shigemoto, Y., Fujii, Y., Shinomiya, K., & Kamei, C. (2004). Participation of histaminergic H1 and noradrenergic alpha 1 receptors in orexin A-induced wakefulness in rats. *Brain Research*, *1023*, 121–125.
Shirasaka, T., Nakazato, M., Matsukura, S., Takasaki, M., & Kannan, H. (1999). Sympathetic and cardiovascular actions of orexins in conscious rats. *The American Journal of Physiology*, *277*, R1780–R1785.
Singh, C., Oikonomou, G., & Prober, D. A. (2015). Norepinephrine is required to promote wakefulness and for hypocretin-induced arousal in zebrafish. *eLife*, *4*, e07000.
Sorooshyari, S., Huerta, R., & de Lecea, L. (2015). A framework for quantitative modeling of neural circuits involved in sleep-to-wake transition. *Frontiers in Neurology*, *6*, 32.
Suzuki, M., Beuckmann, C. T., Shikata, K., Ogura, H., & Sawai, T. (2005). Orexin-A (hypocretin-1) is possibly involved in generation of anxiety-like behavior. *Brain Research*, *1044*, 116–121.
Takahashi, K., et al. (2005). State-dependent effects of orexins on the serotonergic dorsal raphe neurons in the rat. *Regulatory Peptides*, *126*, 43–47.
Takakusaki, K., et al. (2005). Orexinergic projections to the cat midbrain mediate alternation of emotional behavioural states from locomotion to cataplexy. *The Journal of Physiology*, *568*, 1003–1020.
Takahashi, K., Koyama, Y., Kayama, Y., & Yamamoto, M. (2002). Effects of orexin on the laterodorsal tegmental neurones. *Psychiatry and Clinical Neurosciences*, *56*, 335–336.
Takahashi, K., Lin, J. S., & Sakai, K. (2008). Neuronal activity of orexin and non-orexin waking-active neurons during wake-sleep states in the mouse. *Neuroscience*, *153*, 860–870.
Takahashi, N., Okumura, T., Yamada, H., & Kohgo, Y. (1999). Stimulation of gastric acid secretion by centrally administered orexin-A in conscious rats. *Biochemical and Biophysical Research Communications*, *254*, 623–627.
Thakkar, M. M., Ramesh, V., Strecker, R. E., & McCarley, R. W. (2001). Microdialysis perfusion of orexin-A in the basal forebrain increases wakefulness in freely behaving rats. *Archives Italiennes de Biologie*, *139*, 313–328.
Thannickal, T. C., et al. (2000). Reduced number of hypocretin neurons in human narcolepsy. *Neuron*, *27*, 469–474.
Torrealba, F., Yanagisawa, M., & Saper, C. B. (2003). Colocalization of orexin a and glutamate immunoreactivity in axon terminals in the tuberomammillary nucleus in rats. *Neuroscience*, *119*, 1033–1044.
Trivedi, P., Yu, H., MacNeil, D. J., Van der Ploeg, L. H., & Guan, X. M. (1998). Distribution of orexin receptor mRNA in the rat brain. *FEBS Letters*, *438*, 71–75.
Tsujino, N., et al. (2005). Cholecystokinin activates orexin/hypocretin neurons through the cholecystokinin A receptor. *The Journal of Neuroscience*, *25*, 7459–7469.
Tsunematsu, T., et al. (2011). Acute optogenetic silencing of orexin/hypocretin neurons induces slow-wave sleep in mice. *The Journal of Neuroscience*, *31*, 10529–10539.

Tsunematsu, T., et al. (2013). Long-lasting silencing of orexin/hypocretin neurons using archaerhodopsin induces slow-wave sleep in mice. *Behavioural Brain Research*, *255*, 64–74.

Tung, L. W., et al. (2016). Orexins contribute to restraint stress-induced cocaine relapse by endocannabinoid-mediated disinhibition of dopaminergic neurons. *Nature Communications*, *7*, 12199.

Tyree, S. M., & de Lecea, L. (2017a). Optogenetic investigation of arousal circuits. *International Journal of Molecular Sciences*, *18*.

Tyree, S. M., & de Lecea, L. (2017b). Lateral hypothalamic control of the ventral tegmental area: Reward evaluation and the driving of motivated behavior. *Frontiers in Systems Neuroscience*, *11*, 50.

Uramura, K., et al. (2001). Orexin-a activates phospholipase C- and protein kinase C-mediated Ca2+ signaling in dopamine neurons of the ventral tegmental area. *Neuroreport*, *12*, 1885–1889.

Uschakov, A., et al. (2011). Sleep-deprivation regulates alpha-2 adrenergic responses of rat hypocretin/orexin neurons. *PLoS One*, *6*, e16672.

Uschakov, A., Gong, H., McGinty, D., & Szymusiak, R. (2007). Efferent projections from the median preoptic nucleus to sleep- and arousal-regulatory nuclei in the rat brain. *Neuroscience*, *150*, 104–120.

Vittoz, N. M., & Berridge, C. W. (2006). Hypocretin/orexin selectively increases dopamine efflux within the prefrontal cortex: Involvement of the ventral tegmental area. *Neuropsychopharmacology*, *31*, 384–395.

Walling, S. G., Nutt, D. J., Lalies, M. D., & Harley, C. W. (2004). Orexin-A infusion in the locus ceruleus triggers norepinephrine (NE) release and NE-induced long-term potentiation in the dentate gyrus. *The Journal of Neuroscience*, *24*, 7421–7426.

Wang, Q. P., et al. (2005). The orexinergic synaptic innervation of serotonin- and orexin 1-receptor-containing neurons in the dorsal raphe nucleus. *Regulatory Peptides*, *126*, 35–42.

Watanabe, S., Kuwaki, T., Yanagisawa, M., Fukuda, Y., & Shimoyama, M. (2005). Persistent pain and stress activate pain-inhibitory orexin pathways. *Neuroreport*, *16*, 5–8.

Williams, R. H., Alexopoulos, H., Jensen, L. T., Fugger, L., & Burdakov, D. (2008). Adaptive sugar sensors in hypothalamic feeding circuits. *Proceedings of the National Academy of Sciences of the United States of America*, *105*, 11975–11980.

Williams, R. H., Morton, A. J., & Burdakov, D. (2011). Paradoxical function of orexin/hypocretin circuits in a mouse model of Huntington's disease. *Neurobiology of Disease*, *42*, 438–445.

Willie, J. T., et al. (2003). Distinct narcolepsy syndromes in Orexin receptor-2 and Orexin null mice: Molecular genetic dissection of Non-REM and REM sleep regulatory processes. *Neuron*, *38*, 715–730.

Willie, J. T., et al. (2011). Ectopic overexpression of orexin alters sleep/wakefulness states and muscle tone regulation during REM sleep in mice. *Journal of Molecular Neuroscience*, *43*, 155–161.

Winrow, C. J., et al. (2011). Promotion of sleep by suvorexant-a novel dual orexin receptor antagonist. *Journal of Neurogenetics*, *25*, 52–61.

Winrow, C. J., et al. (2012). Pharmacological characterization of MK-6096—A dual orexin receptor antagonist for insomnia. *Neuropharmacology*, *62*, 978–987.

Winsky-Sommerer, R., et al. (2004). Interaction between the corticotropin-releasing factor system and hypocretins (orexins): A novel circuit mediating stress response. *The Journal of Neuroscience*, *24*, 11439–11448.

Xi, M. C., Fung, S. J., Yamuy, J., Morales, F. R., & Chase, M. H. (2002). Induction of active (REM) sleep and motor inhibition by hypocretin in the nucleus pontis oralis of the cat. *Journal of Neurophysiology*, *87*, 2880–2888.

Xi, M. C., Morales, F. R., & Chase, M. H. (2001). Effects on sleep and wakefulness of the injection of hypocretin-1 (orexin-A) into the laterodorsal tegmental nucleus of the cat. *Brain Research*, *901*, 259–264.

Yamanaka, A., et al. (2002). Orexins activate histaminergic neurons via the orexin 2 receptor. *Biochemical and Biophysical Research Communications*, *290*, 1237–1245.

Yamanaka, A., et al. (2003). Hypothalamic orexin neurons regulate arousal according to energy balance in mice. *Neuron*, *38*, 701–713.

Yamanaka, A., Muraki, Y., Tsujino, N., Goto, K., & Sakurai, T. (2003). Regulation of orexin neurons by the monoaminergic and cholinergic systems. *Biochemical and Biophysical Research Communications*, *303*, 120–129.

Yelin-Bekerman, L., et al. (2015). Hypocretin neuron-specific transcriptome profiling identifies the sleep modulator Kcnh4a. *eLife*, *4*, e08638.

Yin, J., & Rosenbaum, D. M. (2017). The human orexin/hypocretin receptor crystal structures. *Current Topics in Behavioral Neurosciences*, *33*, 1–15.

Yin, J., Mobarec, J. C., Kolb, P., & Rosenbaum, D. M. (2015). Crystal structure of the human OX2 orexin receptor bound to the insomnia drug suvorexant. *Nature*, *519*, 247–250.

Yoshida, Y., et al. (2001). Fluctuation of extracellular hypocretin-1 (orexin A) levels in the rat in relation to the light-dark cycle and sleep-wake activities. *The European Journal of Neuroscience*, *14*, 1075–1081.

Yoshida, K., McCormack, S., Espana, R. A., Crocker, A., & Scammell, T. E. (2006). Afferents to the orexin neurons of the rat brain. *Journal of Comparative Neurology*, *494*, 845–861.

Zeitzer, J. M., et al. (2003). Circadian and homeostatic regulation of hypocretin in a primate model: Implications for the consolidation of wakefulness. *The Journal of Neuroscience*, *23*, 3555–3560.

Zhang, W., et al. (2010). Orexin neurons are indispensable for stress-induced thermogenesis in mice. *The Journal of Physiology*, *588*, 4117–4129.

Zhou, W. L., Gao, X. B., & Picciotto, M. R. (2015). Acetylcholine acts through nicotinic receptors to enhance the firing rate of a subset of hypocretin neurons in the mouse hypothalamus through distinct presynaptic and postsynaptic mechanisms. *eNeuro*, *2*, e0052.

CHAPTER 2

Pharmacology of Orexin/ Hypocretin Receptors

Jyrki P. Kukkonen
Department of Physiology, Institute of Biomedicine, Biomedicum Helsinki, and Biochemistry and Cell Biology, Faculty of Medicine, Department of Veterinary Biosciences, Faculty of Veterinary Medicine, University of Helsinki, Helsinki, Finland

ABBREVIATIONS

1- and 2-SORA	OX_1- and OX_2-selective orexin receptor antagonist, respectively
BAT	brown adipose tissue
DORA	dual orexin receptor antagonist
DREADD	designer receptor exclusively activated by designer drugs
GPCR	G protein-coupled receptor
GWAS	genome-wide association study
PET	positron emission tomography
PPO	prepro-orexin

INTRODUCTION

What kind of pharmacological aspects and interests are related to the orexin/hypocretin system? The orexin system is a central player in the regulation of many physiological processes, as described in the other chapters of this book. Therewith the orexinergic signaling may offer a drug target in some disorders of these systems. This is the therapeutic perspective, which indeed is currently actively explored. Genomics-based approaches try to pinpoint some disorders to polymorphisms within the orexin system, or identify which parts of the population would gain from a particular orexinergic therapy. A closely bound interest is the possibility of side effects in the potential orexinergic therapy. In basic and preclinical research perspective we recognize that pharmacology and physiology are intimately related, also as pharmacology will offer tools for discovery of physiological functions. Some orexinergic drugs, especially the early available OX_1-selective antagonist SB-334867, have contributed to important discoveries of orexin physiology.

The Orexin/Hypocretin System
https://doi.org/10.1016/B978-0-12-813751-2.00002-4

However, we still lack essential information about species-specific receptor and receptor subtype functions, especially as concerns the prime target of the therapeutic drugs, humans. Finally, the molecular details of orexin receptor signaling itself are very interesting. While the details of the signaling themselves are not a subject of this review, the roles of potential biased agonism and of orexin receptor heteromerization with other G protein-coupled receptor should be considered in drug discovery.

In the current chapter I shall try to deal with these aspects and the further considerations they give rise to. Only very central orexin receptor ligands are presented; for a reader interested in those hundreds of different ligands synthesized there are plenty of devoted, regularly updated publications (see, e.g., Coleman & Renger, 2010; Roecker, Cox, & Coleman, 2016; Winrow et al., 2018a, 2018b).

WHICH MOLECULES CAN BE AND ARE TARGETED?

Orexins are peptide transmitters and their production is thus primarily dependent on the prepro-orexin (PPO) gene transcription activity, mRNA stability, and translation, as well as peptide packing and posttranslational processing. Clearly, elimination of PPO altogether leads to narcolepsy (Chemelli et al., 1999). One identified human PPO mutation is known to lead to narcolepsy, apparently because of reduced processing and therewith reduced levels of mature orexin peptides (Peyron et al., 2000). Little is known of the enzymes processing orexins (or any) peptides to mature forms, and developing drugs against these is of little interest due to risk for narcolepsy; however, orexinergic overactivity, if found in some disorder, might offer one use for such drugs. It can be assumed that similar enzymes are utilized in the processing of several peptides, so targeting of these would imply low selectivity; however, some peptide-processing enzymes have been successfully targeted in drug discovery (e.g., renin and angiotensin-converting enzyme, dipeptidyl peptidase-4, factor IIa and Xa of the blood coagulation cascade, enkephalinase) and more are actively tested (e.g., kallikrein, neprilysin, matrix metalloproteinases). The packing, trafficking of the vesicles, release, and peptide degradation are also likely to be shared with other transmitters. A targeted inhibition of orexin breakdown would enhance orexinergic tone and thus possibly act as vigilance and cognition enhancer, as long as enough of the endogenous orexin production remains, e.g., in the case of narcolepsy. No such drugs are known.

Orexin receptors, by contrast, have been targeted in numerous projects (see, e.g., Coleman & Renger, 2010; Roecker et al., 2016; Winrow et al., 2018a, 2018b); this clearly has its origins in the experience of G protein-coupled receptors (GPCRs) as "drugable" targets and the assumed good selectivity obtainable. The major approaches have focused on the antagonist development and have mainly been taken in the pharmaceutical industry (see, e.g., Coleman & Renger, 2010; Roecker et al., 2016; Winrow et al., 2018a, 2018b). Only a few reports of agonists or allosteric regulators have been published by mainly academic groups (Cano, Grima, & Benet, 2014; Irukayama-Tomobe et al., 2017; Lee, Reddy, & Kodadek, 2010; Nagahara et al., 2015; Turku et al., 2016; Turku et al., 2017; Yanagisawa, 2010).

OREXIN RECEPTOR COMPETITIVE ANTAGONISTS

Orexin receptors were soon after their discovery targeted by SmithKline Beecham (currently GlaxoSmithKline); as the company representatives also contributed to one of the two original publications (Sakurai et al., 1998), the start clearly occurred even before the first publications. The lead molecule of the first series, SB-334867 (Fig. 1A), was published already in 2000 (Haynes et al., 2000; Smart et al., 2001), two and a half years after the first orexin publications (de Lecea et al., 1998; Sakurai et al., 1998). It is an approximately 100-fold selective antagonist of OX_1 receptors (Porter et al., 2001; Putula & Kukkonen, 2012). Further compounds, such as SB-408124 and SB-674042 (Langmead et al., 2004) (Fig. 1A) display enhanced affinity, but probably not selectivity. The original aim was to find appetite-suppressing drugs, but this was not fulfilled by any of these ligands. However, SB-334867 was soon made available to researchers by the company and then it also became commercially available. For several years, SB-334867 was the only accessible orexin receptor antagonist. It has thus contributed to hundreds of studies and significantly to our knowledge on orexin physiology. However, serious misinterpretations may also have occurred. One such is related to careless dosing. SB-334867 is only 100-fold selective for OX_1, and at high concentrations it thus significantly inhibits OX_2 (and actually also other nondisclosed molecular targets; Porter et al., 2001). A very usual concentration of SB-334867 to use is 10 μM. In ex vivo experiments, orexins are also often used at high concentrations (e.g., 1 μM is usual) and often without determination of the EC_{50} value. Combination of a high antagonist dose/concentration with a high orexin dose/concentration does not tell anything about the receptor subtype

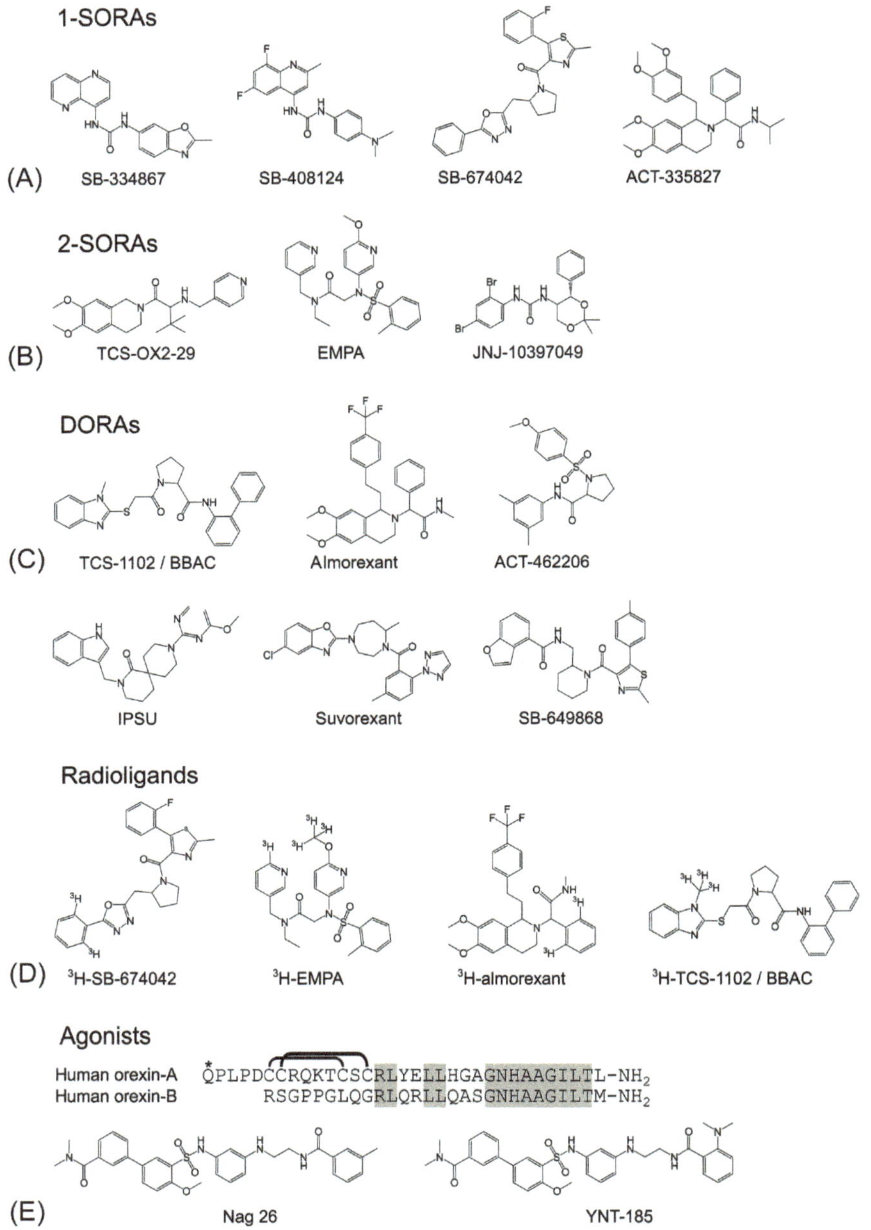

Fig. 1 Some orexin receptor antagonists (A–C), antagonist radioligands (D), and agonists (E). Please observe that the stereochemistry is not shown. For orexin-A, the disulfide bridges are shown, and the C-terminal Q* indicates pyroglutamate. The structures were drawn utilizing ChemDraw Professional 17.0 (PerkinElmer; Waltham, MA, USA).

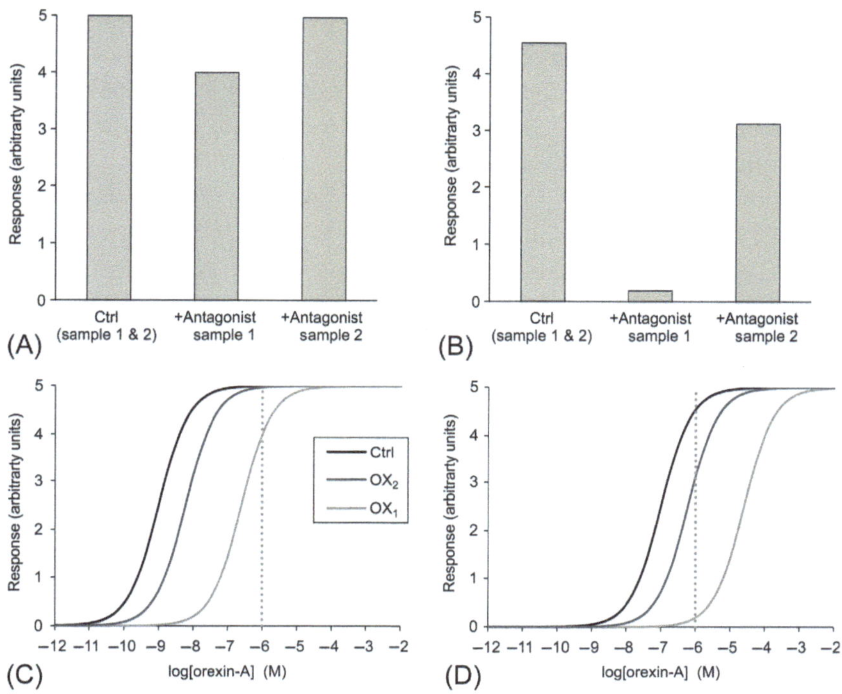

Fig. 2 Simulation of the antagonist (10 μM) effect according to the equation $\text{response} = \frac{[\text{orexin}-\text{A}] \times 100\%}{[\text{orexin}-\text{A}] + \text{EC}_{50}\left(1 + \frac{[10\ \mu\text{M}]}{K_i}\right)}$. The K_i values are taken from one of the original publications with SB-334867 (Porter et al., 2001), i.e., $K_i(\text{OX}_1) = 40$ nM and $K_i(\text{OX}_2) = 2$ μM. In (A) and (B), the responses to 1 μM orexin-A in two different sets of samples (1 and 2) in the presence of 10 μM SB-334867; the control orexin-A response is 100%. In (C) and (D), the full datasets from which the values in (A) and (B), respectively, were taken (1 μM orexin-A is indicated by *dotted gray lines*). In (A) and (C), the EC_{50} of orexin-A is 1 nM, and in (B) and (D), 100 nM.

involved in the response, independent of whether an inhibition is seen or not. I have simulated a few possible conditions in Fig. 2. In A, 1-μM orexin-A and 10-μM SB-334867 are used, with somewhat different responses to the antagonist in different samples. There is no inhibition in sample 2 and also the inhibition in sample 1 is very low, so in the presence of normal variation between samples/experiment, we would draw the conclusion that the response does not really involve OX_1 but is mediated by OX_2. In B, we still have the same orexin-A and SB-334867 concentrations. Sample 1 is now strongly inhibited while also sample 2 is quite clearly partially inhibited. In this case we would be likely to say the response in sample 1 is OX_1 mediated and the response in sample 2 is also partially OX_1 mediated. In subfigures C and D, the full concentration-response curves are revealed. In both cases, sample 1 is OX_1 and sample 2 is OX_2, but the two datasets are different with respect to the EC_{50} of orexin-A. Clearly, we cannot draw too certain conclusions from single concentration experiments!

In addition, SB-334867 was recently described to be unstable in aqueous solutions, and the usual HCl form even in solid state (McElhinny Jr. et al., 2012). Clearly, degraded SB-334867 does not bind orexin receptors, but might it do something else?

Upon the discovery of the role of the orexinergic system in the regulation of sleep and wakefulness, insomnia was soon to become the major indication for discovery of orexin receptor antagonists. Animal studies have supplied partially contradictory evidence of the role of orexin receptor subtypes in sleep regulation: in dogs, naturally occurring, OX_2 receptor-incapacitating mutations appeared to result in strong narcoleptic phenotype (Lin et al., 1999), while in mice, OX_2 knockout gave a milder phenotype, and knockout of both orexin receptor subtypes was needed for strong narcolepsy (Kalogiannis et al., 2011; Mieda et al., 2011; Willie et al., 2003). However, both species supported the more significant role of OX_2 than OX_1 in the regulation of sleep and wakefulness. Both OX_2-selective (2-SORA = OX_2-selective orexin receptor antagonist) and OX_1/OX_2-non-selective antagonists (DORA = dual orexin receptor antagonist) have emerged from (mainly) the company pipelines, and both types have been and are under development for insomnia (even by the same companies). For detailed reviews of the current compounds, please see, e.g. (Coleman & Renger, 2010; Roecker et al., 2016; Winrow et al., 2018a, 2018b). A single compound, suvorexant (Fig. 1C), is on the market in the United States and Japan under the trade name Belsomra® for the indication primary insomnia (both sleep induction and maintenance) and currently in human studies even in Europe. Suvorexant is a DORA: there is no corresponding human information available, and not all suvorexant data have been made public. The orexin receptor expression, as discussed previously, may be different in different species, and sleep architecture definitely is different between human and rodents. It is thus not possible to judge from the currently published data, whether a DORA or a 2-SORA might be more beneficial in insomnia and have less adverse effects.

From purely research point of view we should always test both OX_1- and OX_2-selective antagonists in parallel when determining the receptor subtype responsible for a physiological response. Luckily most current studies do that, though "overdosing" still takes place. Taken the instability of SB-334867, it probably should give place to other ligands altogether. A number of other orexin receptor antagonists, both 1- and 2-SORAs and DORAs, for instance, SB-408124, SB-674042, ACT-335827, TCS-OX2-29, EMPA, JNJ-10397049, TCS-1102/BBAC, almorexant, ACT-462206, IPSU, and SB-649868 (Fig. 1A–C), are commercially available. However, use of these antagonists requires care. First, few of them are highly orexin receptor subtype selective (see, e.g., Winrow et al., 2018a, 2018b), and altogether few compounds are very selective for anything at high concentrations (such as 10 μM). Secondly, for many of the compounds, the affinities have not been properly determined ever, or if they have been, these have not been published. Certainly the company information is not to be blindly trusted. Orexin receptors are luckily highly conserved among mammalian species, but there nevertheless are sequence differences (Tables 1 and 2), which may contribute to affinity differences. Finally, if the compounds are to be used in a whole animal setting, their tissue distribution—and thus actual concentration in situ—may be grossly different, and thus the assumed selectivity does not hold true. Nevertheless, under beneficial conditions, use of a set composed of a single 1-SORA and a single 2-SORA with well-verified affinities would give rather reliable information even without need to obtain orexin concentration–response data as long as both receptor subtypes are not significantly involved or the receptor di-/oligomerization does not change the pharmacology (see later).

OREXIN RECEPTOR AGONISTS

Orexin peptides, native or modified, are the ligands used in nearly all studies. Though different studies present contradictory views on this (Bingham et al., 2001; Deadwyler, Porrino, Siegel, & Hampson, 2007; Fujiki, Yoshida, Ripley, Mignot, & Nishino, 2003; John, Wu, & Siegel, 2000; Kastin & Akerstrom, 1999), it seems likely that they do not show much brain tissue penetration upon peripheral administration. As peptides, it can be assumed that they are not very stable either. Thus the studies have usually resorted to intracerebroventricular administration or injections into specific nuclei. Combination of the native peptides orexin-A and orexin-B (Fig. 1E) has been used to differentiate between the receptor subtypes: according to that, the peptides would be equally potent at OX_2 while orexin-B would be 10–100-fold less potent than orexin-A at OX_1. While this has been seen in several studies, including the original one, with respect to Ca^{2+} elevation

Table 1 Alignment of OX_1 orexin receptor transmembrane helix (TM) sequences of some mammalian species

TM1	Hs	E	W	V	L	I	A	A	Y	V	A	V	F	V	V	A	L	V	G	N	T	L	V	C	L	A	V	W	R	N						
	Cl	E	W	V	L	I	A	A	Y	V	A	V	F	L	V	A	L	V	G	N	T	L	V	C	L	A	V	W	R	N						
	MmA	E	W	V	L	I	A	A	Y	V	A	V	F	L	I	A	L	V	G	N	T	L	V	C	L	A	V	W	R	N						
	Rn	E	W	V	L	I	A	A	Y	V	A	V	F	L	I	A	L	V	G	N	T	L	V	C	L	A	V	W	R	N						
TM2	Hs	T	V	T	N	Y	F	I	V	N	L	S	L	A	D	V	L	V	T	A	I	C	L	P	A	S	L	L	V	D	I	T	E			
	Cl	T	V	T	N	Y	F	I	V	N	L	S	L	A	D	V	L	V	T	A	I	C	L	P	A	S	L	L	V	D	I	T	E			
	MmA	T	V	T	N	Y	F	I	V	N	L	S	L	A	D	V	L	V	T	A	I	C	L	P	A	S	L	L	V	D	I	T	E			
	Rn	T	V	T	N	Y	F	I	V	N	L	S	L	A	D	V	L	V	T	A	I	C	L	P	A	S	L	L	V	D	I	T	E			
TM3	Hs	H	A	L	C	K	V	I	P	Y	L	Q	A	V	S	V	S	V	A	V	L	T	L	S	F	I	A	L	D	R	W	Y	A	I	C	H
	Cl	H	T	L	C	K	V	I	P	Y	L	Q	A	V	S	V	S	V	A	V	L	T	L	S	F	I	A	L	D	R	W	Y	A	I	C	H
	MmA	Q	A	L	C	K	V	I	P	Y	L	Q	A	V	S	V	S	V	A	V	L	T	L	S	F	I	A	L	D	R	W	Y	A	I	C	H
	Rn	H	A	L	C	K	V	I	P	Y	L	Q	A	V	S	V	S	V	A	V	L	T	L	S	F	I	A	L	D	R	W	Y	A	I	C	H
TM4	Hs	T	A	R	R	A	R	G	S	I	L	G[†]	I	W	A	V	S	L	A	I	M	V	P	Q	A	A	V									
	Cl	T	A	R	R	A	R	S	S	I	L	G	I	W	A	V	S	L	A	V	M	V	P	Q	A	A	V									
	MmA	T	A	R	R	A	R	G	S	I	L	G	I	W	A	V	S	L	A	V	M	V	P	Q	A	A	V									
	MmB																				M	V	P	Q	A	A	V									
	Rn	T	A	R	R	A	R	G	S	I	L	G	I	W	A	V	S	L	A	V	M	V	P	Q	A	A	V									
TM5	Hs	D	L	Y	P	K	I	Y	H	S	C	F	F	I	V	T	Y	L	A	P	L	G	L	M	A	M	A	Y	F	Q	I	F	R	K	L	W
	Cl	D	L	Y	P	K	I	Y	H	S	C	F	F	I	V	T	Y	L	A	P	L	G	L	M	A	M	A	Y	F	Q	I	F	R	K	L	W
	MmA	E	L	Y	P	K	I	Y	H	S	C	F	F	I	V	T	Y	L	A	P	L	G	L	M	A	M	A	Y	F	Q	I	F	R	K	L	W
	MmB	E	L	Y	P	K	I	Y	H	S	C	F	F	I	V	T	Y	L	A	P	L	G	L	M	A	M	A	Y	F	Q	I	F	R	K	L	W
	Rn	E	L	Y	P	K	I	Y	H	S	C	F	F	F	V	T	Y	L	A	P	L	G	L	M	G	M	A	Y	F	Q	I	F	R	K	L	W
TM6	Hs	R	A	R	R	K	T	A	K	M	L	M	V	V	L	L	V	F	A	L	C	Y	L	P	I	S	V	L	N	V	L	K	R	V		
	Cl	R	A	R	R	K	T	A	K	M	L	M	V	L	L	L	V	F	A	L	C	Y	L	P	I	S	V	L	N	V	L	K	R	V		
	MmA	R	A	R	R	K	T	A	K	M	L	M	V	V	L	L	V	F	A	L	C	Y	L	P	I	S	V	L	N	V	L	K	R	V		
	MmB	R	A	R	R	K	T	A	K	M	L	M	V	V	L	L	V	F	A	L	C	Y	L	P	I	S	V	L	N	V	L	K	R	V		
	Rn	R	A	R	R	K	T	A	K	M	L	M	V	V	L	L	V	F	A	L	C	Y	L	P	I	S	V	L	N	V	L	K	R	V		
TM7	Hs	E	A	V	Y	A	C	F	T	F	S	H	W	L	V	Y	A	N	S	A	A	N	P	I	I	Y	N	F	L	S						
	Cl	E	A	V	Y	T	C	F	T	F	S	H	W	L	V	Y	A	N	S	A	A	N	P	I	I	Y	N	F	L	S						
	MmA	E	A	V	Y	A	C	F	T	F	S	H	W	L	V	Y	A	N	S	A	A	N	P	I	I	Y	N	F	L	S						
	MmB	E	A	V	Y	A	C	F	T	F	S	H	W	L	V	Y	A	N	S	A	A	N	P	I	I	Y	N	F	L	S						
	Rn	E	A	V	Y	A	C	F	T	F	S	H	W	L	V	Y	A	N	S	A	A	N	P	I	I	Y	N	F	L	S						

The TM domains are defined based on (Karhu, Turku, & Xhaard, 2015; Yin et al., 2016). An additional human polymorphism, 167 Gly/Ser (marked with [†]), has been reported (Kukkonen, Holmqvist, Ammoun, & Åkerman, 2002). Hs, Homo sapiens, entries NP_001516, XP_011539610 and AAH74796; Cl, Canis lupus, entry XP_003638985; MmA, Mus musculus, entries NP_945197, NP_001156499, XP_011249539, XP_011249538, XP_011238820, XP_011238819 and AAI19584; MmB, Mus musculus, entries XP_006537046 and XP_006503124; Rn, Rattus niger, entry NP_037196.

in recombinant Chinese hamster ovary K1 (CHO-K1) cells and also in some other cell lines (Ammoun et al., 2003; Holmqvist, Åkerman, & Kukkonen, 2002; Sakurai et al., 1998), this is not necessarily true in other cell lines or with respect to other responses in CHO-K1 cells (Kukkonen, 2013; Putula, Turunen, Jäntti, Ekholm, & Kukkonen, 2011). The same can be concluded for the reputed "OX_2-selective" synthetic peptide, Ala^{11}, D-Leu^{15}-orexin-B (Putula et al., 2011). Some studies even point out that orexin-B is more potent than orexin-A for some responses. As described later, the orexin receptor signaling is very versatile, and we can assume that the receptor interaction with primary signal transducers lies behind at least a part of the divergent findings. Biased agonism/signaling arises from the differential active receptor conformations induced/stabilized by different agonists, as these conformations differentially activate the primary signal transducers (Kenakin, 2011). Orexin receptors are able to interact with a number of G-proteins and other proteins (see later), and thus biased signaling would be a very likely scenario. Another likely scenario are signal complexes, specifically formed in each cell type and even at different subcellular sites. Such complexes, as well as the entire expression profile of a particular cell, would act in concert with biased signaling. Orexin receptors apparently readily form homo- or heteromeric di- or oligomeric complexes, which might affect the signaling (and pharmacology) (see section "Orexin Receptor Complexes with Other Proteins"). Finally, other concurrent signals received by the cells may affect the orexin responses. I therefore think that in the case of orexin receptors, agonists should not be used for determination of receptor subtype involvement in a particular response. As described previously, there are good antagonists available for this purpose.

Very few small molecule ligands or series have been described. Currently, only one promising series exists (Irukayama-Tomobe et al., 2017; Nagahara et al., 2015), while the other series or compounds described are assumed to be either rather weak or even toxic, as based on published results (or lack thereof) or personal communications (Cano et al., 2014; Turku et al., 2016; Turku et al., 2017; Yanagisawa, 2010). The weakly (compound 26; here called Nag 26) and strongly (compound 31 a.k.a. YNT-185) (Fig. 1E) water-soluble lead molecules of the promising series have been reported to show 70- and 100-fold selectivity, respectively, for OX_2 in the Ca^{2+} elevation assay in CHO cells (Irukayama-Tomobe et al., 2017; Nagahara et al., 2015). Our recent studies with Nag 26 suggest that its OX_2 selectivity is somewhat lower (Rinne et al., 2018). Unfortunately, YNT-185 was found to show weak brain penetration, while Nag 26 has

Table 2 Alignment of OX_2 orexin receptor TM sequences of some mammalian species

TM1	HsA	E	W	V	L	I	A	G	Y	I	I	V	F	V	V	A	L	I	G	N	V	L	V	C	V	A	V	W	K	N						
	HsB	E	W	V	L	I	A	G	Y	I	I	V	F	V	V	A	L	I	G	N	V	L	V	C	V	A	V	W	K	N						
	HsC	E	W	V	L	I	A	G	Y	I	I	V	F	V	V	A	L	I	G	N	V	L	V	C	V	A	V	W	K	N						
	Cl	E	W	V	L	I	A	G	Y	I	I	V	F	V	V	A	L	V	G	N	V	L	V	C	V	A	V	W	K	N						
	MmA	E	W	V	L	I	A	G	Y	I	I	V	F	V	V	A	L	I	G	N	V	L	V	C	V	A	V	W	K	N						
	MmB	E	W	V	L	I	A	G	Y	I	I	V	F	V	V	A	L	I	G	N	V	L	V	C	V	A	V	W	K	N						
	MmC	E	W	V	L	I	A	G	Y	I	I	V	F	V	V	A	L	I	G	N	V	L	V	C	V	A	V	W	K	N						
	RnA	E	W	V	L	I	A	G	Y	I	I	V	F	V	V	A	L	I	G	N	V	L	V	C	V	A	V	W	K	N						
	RnB	E	W	V	L	I	A	G	Y	I	I	V	F	V	V	A	L	I	G	N	V	L	V	C	V	A	V	W	K	N						
TM2	HsA	T	V	T	N	Y	F	I	V	N	L	S	L	A	D	V	L	V	T	I	T	C	L	P	A	T	L	V	V	D	I	T	E			
	HsB	T	V	T	N	Y	F	I	V	N	L	S	L	A	D	V	L	V	T	I	T	C	L	P	A	T	L	V	V	D	I	T	E			
	HsC	T	V	T	N	Y	F	I	V	N	L	S	L	A	D	V	L	V	T	I	T	C	L	P	A	T	L	V	V	D	I	T	E			
	HsD	T	V	T	N	Y	F	I	V	N	L	S	L	A	D	V	L	V	T	I	T	C	L	P	A	T	L	V	V	D	I	T	E			
	Cl	T	V	T	N	Y	F	I	V	N	L	S	L	A	D	V	L	V	T	I	T	C	L	P	A	T	L	V	V	D	I	T	E			
	MmA	T	V	T	N	Y	F	I	V	N	L	S	L	A	D	V	L	V	T	I	T	C	L	P	A	T	L	V	V	D	I	T	E			
	MmB	T	V	T	N	Y	F	I	V	N	L	S	L	A	D	V	L	V	T	I	T	C	L	P	A	T	L	V	V	D	I	T	E			
	MmC	T	V	T	N	Y	F	I	V	N	L	S	L	A	D	V	L	V	T	I	T	C	L	P	A	T	L	V	V	D	I	T	E			
	RnA	T	V	T	N	Y	F	I	V	N	L	S	L	A	D	V	L	V	T	I	T	C	L	P	A	T	L	V	V	D	I	T	E			
	RnB	T	V	T	N	Y	F	I	V	N	L	S	L	A	D	V	L	V	T	I	T	C	L	P	A	T	L	V	V	D	I	T	E			
TM3	HsA	Q	S	L	C	K	V	I	P	Y	L	Q	T	V	S	V	S	V	S	V	L	T	L	S	C	I	A	L	D	R	W	Y	A	I	C	H
	HsB	Q	S	L	C	K	V	I	P	Y	L	Q	T	V	S	V	S	V	S	V	L	T	L	S	C	I	A	L	D	R	W	Y	A	I	C	H
	HsC	Q	S	L	C	K	V	I	P	Y	L	Q	T	V	S	V	S	V	S	V	L	T	L	S	C	I	A	L	D	R	W	Y	A	I	C	H
	HsD	Q	S	L	C	K	V	I	P	Y	L	Q	T	V	S	V	S	V	S	V	L	T	L	S	C	I	A	L	D	R	W	Y	A	I	C	H
	Cl	Q	S	L	C	K	V	I	P	Y	L	Q	T	V	S	V	S	V	S	V	L	T	L	S	C	I	A	L	D	R	W	Y	A	I	C	H
	MmA	Q	S	L	C	K	V	I	P	Y	L	Q	T	V	S	V	S	V	S	V	L	T	L	S	C	I	A	L	D	R	W	Y	A	I	C	H
	MmB	Q	S	L	C	K	V	I	P	Y	L	Q	T	V	S	V	S	V	S	V	L	T	L	S	C	I	A	L	D	R	W	Y	A	I	C	H
	MmC	Q	S	L	C	K	V	I	P	Y	L	Q	T	V	S	V	S	V	S	V	L	T	L	S	C	I	A	L	D	R	W	Y	A	I	C	H
	RnA	Q	S	L	C	K	V	I	P	Y	L	Q	T	V	S	V	S	V	S	V	L	T	L	S	C	I	A	L	D	R	W	Y	A	I	C	H
	RnB	Q	S	L	C	K	V	I	P	Y	L	Q	T	V	S	V	S	V	S	V	L	T	L	S	C	I	A	L	D	R	W	Y	A	I	C	H
TM4	HsA	T	A	K	R	A	R	N	S	I	V	I	I	W	I	V	S	C	I	I	M	I	P	Q	A	I	V									
	HsB	T	A	K	R	A	R	N	S	I	V	I	I	W	I	V	S	C	I	I	M	I	P	Q	A	I	V									
	HsC	T	A	K	R	A	R	N	S	I	V	I	I	W	I	V	S	C	I	I	M	I	P	Q	A	I	V									
	HsD	T	A	K	R	A	R	N	S	I	V	I	I	W	I	V	S	C	I	I	M	I	P	Q	A	I	V									
	Cl	T	A	K	R	A	R	N	S	I	V	I	I	W	I	V	S	C	I	I	M	I	P	Q	A	I	V									
	MmA	T	A	K	R	A	R	N	S	I	V	V	I	W	I	V	S	C	I	I	M	I	P	Q	A	I	V									
	MmB	T	A	K	R	A	R	N	S	I	V	V	I	W	I	V	S	C	I	I	M	I	P	Q	A	I	V									
	MmC	T	A	K	R	A	R	N	S	I	V	V	I	W	I	V	S	C	I	I	M	I	P	Q	A	I	V									
	RnA	T	A	K	R	A	R	N	S	I	V	V	I	W	I	V	S	C	I	I	M	I	P	Q	A	I	V									
	RnB	T	A	K	R	A	R	N	S	I	V	V	I	W	I	V	S	C	I	I	M	I	P	Q	A	I	V									

TM5	HsA	E	I	Y	P	K	M	Y	H	I	C	F	F	L	V	T	Y	M	A	P	L	C	L	M	V	L	A	Y	L	Q	I	F	R	K	L	W
	HsB	E	I	Y	P	K	M	Y	H	I	C	F	F	L	V	T	Y	M	A	P	L	C	L	M	V	L	A	Y	L	Q	I	F	R	K	L	W
	HsC	E	I	Y	P	K	M	Y	H	I	C	F	F	L	V	T	Y	M	A	P	L	C	L	M	V	L	A	Y	L	Q	I	F	R	K	L	W
	HsD	E	I	Y	P	K	M	Y	H	I	C	F	F	L	V	T	Y	M	A	P	L	C	L	M	V	L	A	Y	L	Q	I	F	R	K	L	W
	Cl	E	I	Y	P	K	M	Y	H	I	C	F	F	L	V	T	Y	M	A	P	L	C	L	M	V	L	A	Y	L	Q	I	F	R	K	L	W
	MmA	E	V	Y	P	K	M	Y	H	I	C	F	F	L	V	T	Y	M	A	P	L	C	L	M	I	L	A	Y	L	Q	I	F	R	K	L	W
	MmB	E	V	Y	P	K	M	Y	H	I	C	F	F	L	V	T	Y	M	A	P	L	C	L	M	V	L	A	Y	L	Q	I	F	R	K	L	W
	MmC	E	V	Y	P	K	M	Y	H	I	C	F	F	L	V	T	Y	M	A	P	L	C	L	M	I	L	A	Y	L	Q	I	F	R	K	L	W
	RnA	E	V	Y	P	K	M	Y	H	I	C	F	F	L	V	T	Y	M	A	P	L	C	L	M	V	L	A	Y	L	Q	I	F	R	K	L	W
	RnB	E	V	Y	P	K	M	Y	H	I	C	F	F	L	V	T	Y	M	A	P	L	C	L	M	V	L	A	Y	L	Q	I	F	R	K	L	W
TM6	HsA	R	A	R	R	K	T	A	R	M	L	M	I	V	L	L	V	F	A	I	C	Y	L	P	I	S	I	L	N	V	L	K	R	V		
	HsB	R	A	R	R	K	T	A	R	M	L	M	V	V	L	L	V	F	A	I	C	Y	L	P	I	S	I	L	N	V	L	K	R	V		
	HsD	R	A	R	R	K	T	A	R	M	L	M	I	V	L	L	V	F	A	I	C	Y	L	P	I	S	I	L	N	V	L	K	R	V		
	Cl	R	A	R	R	K	T	A	R	M	L	M	V	V	L	L	V	F	A	I	C	Y	L	P	I	S	I	L	N	V	L	K	R	V		
	MmA	R	A	R	R	K	T	A	R	M	L	M	V	V	L	L	V	F	A	I	C	Y	L	P	I	S	I	L	N	V	L	K	R	V		
	MmB	R	A	R	R	K	T	A	R	M	L	M	V	V	L	L	V	F	A	I	C	Y	L	P	I	S	I	L	N	V	L	K	R	V		
	MmC	R	A	R	R	K	T	A	R	M	L	M	V	V	L	L	V	F	A	I	C	Y	L	P	I	S	I	L	N	V	L	K	R	V		
	RnA	R	A	R	R	K	T	A	R	M	L	M	V	V	L	L	V	F	A	I	C	Y	L	P	I	S	I	L	N	V	L	K	R	V		
	RnB	R	A	R	R	K	T	A	R	M	L	M	V	V	L	L	V	F	A	I	C	Y	L	P	I	S	I	L	N	V	L	K	R	V		
TM7	HsA	E	T	V	Y	A	W	F	T	F	S	H	W	L	V	Y	A	N	S	A	A	N	P	I	I	Y	N	F	L	S						
	HsB	E	T	V	Y	A	W	F	T	F	S	H	W	L	V	Y	A	N	S	A	A	N	P	I	I	Y	N	F	L	S						
	HsD	E	T	V	Y	A	W	F	T	F	S	H	W	L	V	Y	A	N	S	A	A	N	P	I	I	Y	N	F	L	S						
	Cl	E	T	V	Y	A	W	F	T	F	S	H	W	L	V	Y	A	N	S	A	A	N	P	I	I	Y	N	F	L	S						
	MmA	E	T	V	Y	A	W	F	T	F	S	H	W	L	V	Y	A	N	S	A	A	N	P	I	I	Y	N	F	L	S						
	MmB	E	T	V	Y	A	W	F	T	F	S	H	W	L	V	Y	A	N	S	A	A	N	P	I	I	Y	N	F	L	S						
	MmC	E	T	V	Y	A	W	F	T	F	S	H	W	L	V	Y	A	N	S	A	A	N	P	I	I	Y	N	F	L	S						
	RnA	E	T	V	Y	A	W	F	T	F	S	H	W	L	V	Y	A	N	S	A	A	N	P	I	I	Y	N	F	L	S						
	RnB	E	T	V	Y	A	W	F	T	F	S	H	W	L	V	Y	A	N	S	A	A	N	P	I	I	Y	N	F	L	S						

The TM domains are defined based on (Karhu et al., 2015; Yin et al., 2016; Yin, Mobarec, Kolb, & Rosenbaum, 2015). Please see (Thompson et al., 2014) for details. HsA, Homo sapiens, entries NP_001517, AGP04116 and EAX04440; HsB, Homo sapiens, entry AAC39602; HsC, Homo sapiens, entry AAGP04117; HsD, Homo sapiens, entry XP_011512844; Cl, Canis lupus, entries NP_001002933, AAD49333 and Q9TUP7; MmA, Mus musculus, entries NP_945200, EDL26342, AAI40958, AAR01328 and AAR11293; MmB, Mus musculus, entry EDL26341; MmC, Mus musculus, entries XP_011241068, AAR01327 and AAR11294; RnA, Rattus niger, entries NP_037206, EDL77772 and AAC40042; RnB, Rattus niger, entry XP_008764437.

not been tested (Irukayama-Tomobe et al., 2017). No compound of this series is commercially available. Thus the researchers are currently left without any reasonable substitute for orexin peptides.

Allosteric orexin receptor activators might be yet another potential type of drug. Due to the current, rather qualitative than quantitative methods for orexin measurements, it is unknown what the factual orexin levels in the brain (or even in the cerebrospinal fluid) are under different conditions. It is actually even uncertain whether type 1 narcolepsy is associated with a total loss of orexinergic activity. Thus in different potential disorders with reduced orexinergic signaling, orexin receptor positive allosteric agonists might be very useful. Currently there is a single study suggesting such molecules (Lee et al., 2010), but this study lacks obvious controls and there has been no further development.

LABELED OREXIN RECEPTOR LIGANDS

For years, ^{125}I-orexin-A (label at Tyr^{17}) was the only commercially available radioligand, which is very unfortunate since it possesses most major disadvantageous properties for a radioligand (Kukkonen, 2012). Pharmaceutical companies have published studies utilizing custom-labeled antagonist radioligands (^{3}H-SB-674042, ^{3}H-EMPA, ^{3}H-almorexant, and ^{3}H-TCS-1102/BBAC) (Callander et al., 2013; Langmead et al., 2004; Malherbe et al., 2009; Malherbe et al., 2010; Malherbe, Borroni, Pinard, Wettstein, & Knoflach, 2009) (Fig. 1D), but these have not been commercially available. However, a few such ligands, i.e., ^{3}H-SB-674042, ^{3}H-EMPA, and ^{3}H-almorexant have recently become available. Some characterization of these ligands is available (Langmead et al., 2004; Malherbe, Borroni, Gobbi, et al., 2009), but these suffer from significant shortcomings, such as determination of nonspecific ^{3}H-EMPA binding using an excess of nonlabeled EMPA and some unexplained details in the experimental protocols. Our own unpublished results suggest some anomalous properties for one of these ligands. It would be of wide interest that proper characterization of these ligands, including shell life, was performed and published by an independent laboratory. Currently, another thing limiting the use of these radioligands is their high price.

In-house-made fluorescently labeled orexin-A (label likely in Lys^{10}), as well as the commercially available N-terminally labeled orexin-A, have been utilized in a few studies (Darker et al., 2001; Evans et al., 2001; Karteris,

Randeva, Grammatopoulos, Jaffe, & Hillhouse, 2001). Also N-terminal fluorescently labeled orexin-B and Ala^{11}, D-Leu^{15}-orexin-B, as well as biotinylated orexin peptides are commercially available, but have, to my knowledge, not been used in any published studies. No inherently fluorescent or fluorescently labeled small molecule orexin receptor ligands have been described. Even the utilized, labeled orexin-A variants have not been characterized, and thus their suitability for pharmacological binding studies is essentially unknown. However, our unpublished studies (Kukkonen) indicate that addition in the N-terminus of orexin-A is tolerated without any loss of potency. Determination of IC_{50} values for some truncated orexin receptor peptides has been performed by flow cytometry utilizing competition against fluorescent orexin-A (Darker et al., 2001), and one company sells a time-resolved fluorescence/Förster energy transfer (TR-FRET)-based binding assay utilizing fluorescently labeled orexin-A. The commercially available labeled orexin peptides are very expensive, but more reasonable prices are obtained for custom-synthesized peptides. Another potential use for the fluorescent orexin receptor ligands would be orexin receptor identification in tissues, as the antibodies against orexin receptors are generally not good (Kukkonen, 2012, 2013; similar more recent own unpublished findings).

Some attempts to generate PET (positron emission tomography) ligands toward orexin receptors have been made (Gao, Wang, & Zheng, 2016; Liu et al., 2012; Oi et al., 2013; Wang et al., 2013; Wang et al., 2013). A PET ligand would in principle allow orexin receptor visualization in the brain and other parts of the body, but as it competes with the endogenous orexin peptides, it would rather indicate the orexinergic tonus in the brain. In narcoleptics with low or absent orexin peptides, the receptor distribution as such could be visualized to determine the feasibility of orexinergic therapy. Although the PET ligands are based on previously identified reasonably selective orexin receptor antagonists, the step to PET ligand is huge, as a PET ligand should possess the additional properties of good brain penetration, high affinity and low nonspecific binding, and good retention and biostability. In addition, the ligand needs to be labeled and purified very rapidly due to the short half-life of the PET radionuclides. The approach chosen in the orexin receptor studies has been to directly utilize a known antagonist with one of the carbons as ^{11}C or structures similar to known series. Unfortunately, none of the ligands tested so far has been directly suitable, but some series may show promise.

OTHER TOOLS FOR OREXIN RECEPTOR RESEARCH

As indicated previously, a good number of orexin receptor antagonists have been developed and quite a few are also available, while there is a shortage of agonists. A number of genetic or genomic tools have also been produced, including orexinergic neurons expressing green fluorescent protein; orexin peptide and receptor knockout mice; mice in which orexin neurons are postnatally destroyed; and viral vectors for orexin overexpression or manipulation of the knock-in animals (Kukkonen, 2013; Tabuchi et al., 2014). These have been extremely useful in elucidating the physiological functions of orexins, as described in (Kukkonen, 2013). Specific expression in orexinergic neurons is possible utilizing the proximal PPO promoter (Li, Gao, Sakurai, & van den Pol, 2002; Matsuki et al., 2009; Sakurai et al., 1999).

One additional approach has been genetic manipulation of orexinergic neurons or neurons up- or downstream of them. The most widely used technique is transient expression of *optogenetic* constructs that can be activated by light via light guides to stimulate or inhibit the neurons (depending on the construct type); some of this work has been reviewed in (de Lecea, 2015; Ono & Yamanaka, 2017). The pharmacological approach is to express DREADDs (designer receptors exclusively activated by designer drugs) that—again depending on the receptor type—can be activated to stimulate or inhibit the neurons utilizing the drugs administered. This approach is also called *chemogenetics*, following the convention from optogenetics. It has also been wrongfully called pharmacogenetics, but this term has for long been reserved to quite a different approach. It is clear that optogenetics can offer much higher spatial and temporal resolution for the manipulation and thus also for the conclusions, but also DREADDs can offer important information, and that also often more easily. The findings with DREADDs concerning orexins (Bunney, Zink, Holm, Billington, & Kotz, 2017; Goforth, Leinninger, Patterson, Satin, & Myers Jr., 2014; Grafe et al., 2017; Hasegawa, Yanagisawa, Sakurai, & Mieda, 2014; Sasaki et al., 2011; Soya et al., 2017; Zink, Bunney, Holm, Billington, & Kotz, 2017) have thus far been in agreement with the findings using other methods, but may have to be verified. Specifically, a few aspects have to be considered. Firstly, the setting with expression of a construct under a rather general promoter, combined with systemic drugs administration, does not imply much specificity. Secondly, the normal regulation of the neurons involved is not fully known, and therefore a sustained stimulation, as accomplished with a drug, may lead to a cellular effect different from the one expected. Thirdly, it has recently been shown that the most popular "designer drug" used with DREADDs, clozapine-*N*-oxide, is not the physiological ligand of the DREADDs whereas its breakdown product, clozapine, is (Gomez et al., 2017). The metabolism of clozapine-*N*-oxide to clozapine is fast enough to produce substantial amounts of clozapine (Jann, Lam, & Chang, 1994; MacLaren et al., 2016), especially if high doses of clozapine-*N*-oxide are used, as is the case in some DREADD studies. Clozapine is by no means selective for DREADDs and can cause off-target effects at higher doses. Clozapine has been shown to activate—directly or indirectly—orexinergic neurons in rat (Fadel, Bubser, & Deutch, 2002), which further stresses the importance of critical view and proper experimental controls.

WHICH PHYSIOLOGICAL PROCESSES AND DISORDERS MIGHT BE TARGETED BY OREXINERGIC THERAPEUTICS?

Before taking a look at the potential indications, we must recognize a few limitations. As already mentioned earlier, the grand majority of the ideas of the physiological and pathophysiological role of orexins comes from rodents, while nearly nothing is known about humans, the usual focus of the prime interest. However, there may be some interesting, though currently rather vague, indications from studies of human orexin receptor polymorphisms (Thompson, Xhaard, Sakurai, Rainero, & Kukkonen, 2014). Another limitation are the problems with the properties and the use of the ligands, as discussed previously. The third limitation is the inability to perform quantitative orexin measurements in the CNS or blood (Kukkonen, 2012; Nishino, 2005), which precludes identification of disorders involving changes in orexin levels.

Sleep and Wakefulness

The physiological orexin functions are discussed in much detail in other reviews (e.g., Kukkonen, 2013), and thus these are only briefly mentioned here. The two clearly most significant processes are regulation of (a) sleep and wakefulness and (b) metabolism. The role of orexins in the regulation of wakefulness and sleep stages is most dramatically illustrated in type 1 narcolepsy. However, also other disorders of sleep and wakefulness may be treatable by orexinergic therapeutics. The only currently therapeutically used orexinergic drug, suvorexant, is a rather nonselective DORA, approved in the United States and Japan for use in primary insomnia. There are other DORAs and 2-SORAs in the development pipelines for the same indication (see, e.g., Roecker et al., 2016). As indicated by the animal studies (see previously), OX_2 receptor is more elementarily linked to regulation of sleep and wakefulness in mice and dogs. Clearly, orexin receptor blockers should reduce vigilance and induce sleep, while orexin receptor agonists would do the opposite, and be beneficial in, e.g., narcolepsy.

Appetite and Metabolism

Currently, the role of orexins in the appetite stimulation is considered less significant as compared to many other transmitters and actually redundant. In rodents, centrally administered orexins elevate the metabolic rate, most likely via activation of sympathetic brain stem nuclei (Lubkin & Stricker-Krongrad, 1998). Orexin peptide or orexin neuron deficiency gives different phenotypes in different backgrounds and schemes, but in general feeding is lower and obesity greater (Hara et al., 2001; Willie, Chemelli, Sinton, & Yanagisawa, 2001; Hara, Yanagisawa, & Sakurai, 2005; Fujiki et al., 2006; Ramanathan & Siegel, 2014; see also discussion in Kukkonen, 2012). In human narcoleptics, higher prevalence of overweight and obesity are typical clinical findings (Dahmen, Bierbrauer, & Kasten, 2001; Kok et al., 2003; Kotagal, Krahn, & Slocumb, 2004; Ponziani et al., 2016; Schuld, Hebebrand, Geller, & Pollmacher, 2000), but this is not associated with increased resting metabolic rate (Dahmen, Tonn, Messroghli, Ghezel-Ahmadi, & Engel, 2009; Fronczek et al., 2008). At the current state of knowledge, these findings cannot be directly linked to orexins. Rats and mice, unlike human or most other mammals, retain significant active brown adipose tissue (BAT) even in adulthood, and orexins are a significant factor activating BAT via brain stem nuclei (e.g., raphé pallidus; Oldfield et al., 2002; Berthoud, Patterson, Sutton, Morrison, & Zheng, 2005; Zheng, Patterson, & Berthoud, 2005; Tupone, Madden, Cano, & Morrison, 2011; Kukkonen, 2013). Ectopic overexpression of orexin peptides in the CNS of rats protects against obesity under high-calorie diet by activating BAT (Funato et al., 2009). Interestingly, placentally produced orexin peptides are also needed for proper development of BAT in mice (OX_1-mediated response) (Sellayah, Bharaj, & Sikder, 2011; Sellayah & Sikder, 2012).

Rodent studies suggest orexin regulation of metabolism via hormonal mediators. The findings are only few and disparate, but orexins may regulate adrenocorticotropin (ACTH) release via central pathways (Al-Barazanji, Wilson, Baker, Jessop, & Harbuz, 2001; Brunton & Russell, 2003; Kuru et al., 2000; Samson & Taylor, 2001; Samson, Taylor, Follwell, & Ferguson,

2002). In human narcoleptics, the diurnal rhythmicity of ACTH and cortisol secretion is blunted (Kok et al., 2002), but in rat, almorexant does not affect ACTH or corticosterone release under nonstressful conditions (Steiner et al., 2012). Glucocorticoid release from adrenal cortex of both humans and rodents can be directly stimulated by orexins (Malendowicz et al., 2001; Malendowicz, Tortorella, & Nussdorfer, 1999; Mazzocchi, Malendowicz, Gottardo, Aragona, & Nussdorfer, 2001) (OX_2-mediated response?). Some studies report peripheral actions of orexins on insulin release in rats (see, e.g., Nowak et al., 2005; Nowak, Mackowiak, Switonska, Fabis, & Malendowicz, 2000; Park et al., 2015), but whether this is a direct or an indirect effect is unknown. Orexin-A regulates growth hormone secretion in sheep (Chen & Xu, 2003), and the diurnal rhythm of growth hormone secretion is altered in human narcoleptics (Overeem et al., 2003). Growth hormone has been suggested to be essential for orexin-induced feeding in rat (Alvarez-Crespo et al., 2013).

In general, the regulation of metabolism by orexin receptor-targeting drugs is an interesting possibility, but the findings, especially due to the human vs. rodent differences, do not allow clear conclusions of the putative indications or side effects emerging upon orexinergic therapy. The peripheral effects of orexins are altogether a mystery—as peripheral orexin expression is limited (Kukkonen, 2013)—but the expression of orexin receptors in peripheral tissues makes them reasonable drug targets. Something about the role of central and peripheral orexin receptors in metabolism will undoubtedly be revealed upon the use of suvorexant as hypnotic. Recent rodent studies indicate beneficial role of central and peripheral orexin receptor stimulation in type 2 diabetes and its pathogenesis (Park et al., 2015; Tsuneki et al., 2015).

Cancer

Activation of either orexin receptor subtype is capable of inducing programmed cell death in a variety of cancer-derived cell lines and other immortal cell lines (Alexandre et al., 2014; Ammoun et al., 2003; Ammoun, Lindholm, Wootz, Åkerman, & Kukkonen, 2006; Rouet-Benzineb et al., 2004; Voisin, Firar, Avondo, & Laburthe, 2006). There are a few different mechanisms suggested (Ammoun et al., 2006; El Firar et al., 2009), but unfortunately the picture is not complete or concurrent, and the studies have not continued since these publications. The clinically most interesting studies show OX_1 receptor expression and induction of programmed cell death in human primary colon carcinoma and its liver metastases (Voisin et al., 2011). Orexin-A is even effective in cells resistant to 5-fluorouracil. By contrast, the corresponding normal tissues are devoid of orexin receptor expression and thus also of the apoptotic response (Voisin et al., 2011).

These studies suggest that either orexin receptor subtype might be a reasonable target in cancer. A membrane drug target would evade some of the typical resistance mechanisms of chemotherapeutic drugs, including reduced cellular uptake and increased extrusion (Holohan, Van Schaeybroeck, Longley, & Johnston, 2013). However, the studies identify no role of orexin receptors in the cancer cell phenotype, and if there is none significant, the cells might even more easily adapt to the drug. It also is unclear which cancer cell types—even initially—express orexin receptors. The models, in which the cancer cell death has been observed, are limited. Nevertheless, it would be highly interesting to test any small molecule orexin receptor ligand in relevant cancer models. I still feel that the molecular mechanism of orexin-mediated programmed cell death would be much worthwhile to investigate, as it is possible that such studies would identify an intracellular cancer drug target that could be utilized even in the cells not expressing orexin receptors.

In the absence of continued "hardcore" studies based on the previous findings, we should consider possible problems. What if orexin receptor expression has some autoregulatory role in the cancer cells, and blocking these receptors by, e.g., suvorexant may actually promote cancer growth? Orexin receptor activation is able to stimulate cell proliferation and survival in other cell types (Kim et al., 2015; Skrzypski et al., 2012; Spinazzi et al., 2005; Tang et al., 2008; Wang et al., 2014; Zwirska-Korczala et al., 2007), which would make it potentially counterproductive in cancer therapy. Orexin receptor-mediated cell death has not been assessed for primary neurons; certainly we would not like to risk neuronal cell death during cancer therapy.

Other Disorders

Orexin receptor mutations/polymorphisms have been identified in human populations, and both single findings and targeted studies have suggested, for a few of these, links with disorder phenotypes. As reviewed in (Thompson et al., 2014), the disorders include Tourette's syndrome with attention-deficit hyperactivity disorder (ADHD), cluster headache, migraine, polydipsia-hyponatremia in schizophrenia, depression, and panic disorder.

The links are usually weak or undefined, and most have not been analyzed in genome-wide association studies (GWASs). Most of the polymorphisms have not been analyzed with respect to other diseases than narcolepsy. The risks, when found, have been very moderate. There are no conclusive studies identifying molecular differences between orexin receptor polymorphic variants, and no transgenic animals with the variant receptors have been produced; therefore it cannot be excluded that risks observed are actually related to other cosegregating gene variants. However, there are some findings from animal experiments that support the role of orexinergic mechanisms. Several studies identify orexinergic mechanism in anxiety and depression (Abbas et al., 2015; Avolio, Biasone, Mele, & Alo, 2014; Bonaventure et al., 2017; Jalewa, Wong-Lin, McGinnity, Prasad, & Holscher, 2014; Johnson et al., 2010; Johnson et al., 2012; Khalil & Fendt, 2017; Li et al., 2010; Lungwitz et al., 2012; Mikrouli, Wortwein, Soylu, Mathe, & Petersen, 2011; Scott et al., 2011; Suzuki, Beuckmann, Shikata, Ogura, & Sawai, 2005). OX_1 receptor block has been suggested as a possible therapeutic strategy in anxiety (Abbas et al., 2015; Bonaventure et al., 2017), and it also facilitates fear extinction (Flores et al., 2014). A DORA has been shown to reduce the efficacy of migraine-inducing stimulus (Hoffmann et al., 2015). Orexin receptors have been suggested to decrease or increase drinking (water intake) depending on the stimulus (Karasawa, Yakabi, Wang, & Tache, 2014; Kunii et al., 1999; Milella et al., 2010).

A confounding link is that insomnia itself indirectly increases (or at least is assumed to increase) the risk for depression and metabolic disorder, and thus manipulation of sleep through the orexinergic system may affect these disorders via improvement of the sleep rather than directly. Interesting studies along these lines have been published (Nakamura & Nagamine, 2017; Tsuneki et al., 2016); however, we must be careful not to construct possibly artificial causal relationships.

Several animal studies have also recognized that central orexins reduce seizure threshold while SB-334867 elevates the seizure threshold (Akbari et al., 2014; Erken et al., 2012; Kordi Jaz et al., 2017; Kortunay et al., 2012; Socala, Szuster-Ciesielska, & Wlaz, 2016). However, it is not quite clear whether SB-334867 is OX_1 selective in this context.

In a recent study, genome-wide association approach was used to identify gene variants in response to therapy in human heart failure (Perez et al., 2015). The studies identified a single nucleotide polymorphism (SNP), rs7767652, ~3 kB upstream of the OX_2 gene (*Hcrtr2*)—in this case defining a C-to-T-exchange—which correlated with a poor response to therapy. The authors assumed this variant to reduce binding of the transcription factor TCF4, possibly reducing the receptor expression. Based on further testing in mice, the authors suggested that elimination of OX_2 induces cardiac dysfunction while orexin-A improves cardiac function under conditions mimicking heart failure, and thus orexin-receptor activating therapy would be beneficial in heart failure (Perez et al., 2015). In contrast, central orexins elevate blood pressure and heart rate, and orexin receptor inhibition may be protective even in "spontaneous" hypertension (see, e.g., Samson, Gosnell, Chang, Resch, & Murphy, 1999; Lin, Matsumura, Tsuchihashi, Abe, & Iida, 2002; Machado, Bonagamba, Dun, Kwok, & Dun, 2002; Smith, Connolly, & Ferguson, 2002; Huang, Dai, Lee, Chiou, & Hwang, 2010; Li, Hindmarch, Nattie, & Paton, 2013; Jackson et al., 2016). Therefore the role of orexins in cardiovascular function seems dual: Based on this, a peripheral, poorly brain-penetrant orexin receptor agonist (Irukayama-Tomobe et al., 2017) might be beneficial in heart failure, while a mainly centrally acting antagonist could ameliorate hypertension. However, these are rather early days. It also seems premature to suggest that orexins would be The Link between sleep disturbances and cardiovascular disorders.

Addiction and other motivation-, reward-, and reinforcement-linked processes have been associated with orexins, both with respect to drugs of abuse as well as to palatable food and foodstuffs. Orexins stimulate and potentiate glutamatergic stimulation of ventral tegmental area dopaminergic neurons projecting to, for instance, nucleus accumbens (see, e.g., Nakamura et al., 2000; Uramura et al., 2001; Korotkova, Sergeeva, Eriksson, Haas, & Brown, 2003; Borgland, Taha, Sarti, Fields, & Bonci, 2006; Borgland, Storm, & Bonci, 2008; Baimel, Lau, Qiao, & Borgland, 2017; reviewed in Baimel et al., 2015). Orexins have been implicated in self-administration of some substances but not of others or of specific substances only under certain conditions (Aston-Jones et al., 2010; Baimel et al., 2015; Baimel & Borgland, 2017; Cason et al., 2010; James, Mahler, Moorman, & Aston-Jones, 2017). Whether orexin receptor stimulation itself is rewarding and reinforcing, is not known, due to the lack of small molecule orexin receptor activators. However, it definitely contributes to other processes, such as contextual memory related to abuse and drug relapse, for instance, in stress (Boutrel et al., 2005; Harris, Wimmer, & Aston-Jones, 2005). Orexin receptor inhibition may also inhibit drug-induced plasticity upon development of addiction, drug seeking, and relapse (Winrow et al., 2010).While mainly OX_2 receptors have been associated with orexin actions in ventral tegmental area, drug seeking and other such systemic responses have been inhibitable with the OX_1-selective antagonist SB-334867 (Aston-Jones et al., 2010; Borgland et al., 2008; Boutrel et al., 2005; Cason et al., 2010; Gozzi et al., 2011; Harris et al., 2005; Malherbe, Borroni, Pinard, et al., 2009; Navarro et al., 2015; Smith, See, & Aston-Jones, 2009); thus orexin actions may not only take place via ventral tegmental area, if we are to trust the selectivity of SB-334867 (see section "Orexin Receptor Competitive Antagonists"). It is thinkable that orexin receptor antagonists, maybe especially OX_1 antagonists, would be useful in some types of addiction. However, thus far the evidence is solely from rodents.

PHARMACOLOGY OF OREXIN RECEPTOR SIGNALING

We have reviewed orexin receptor signaling in detail in recent publications (Kukkonen, 2012, 2013; Kukkonen, 2017; Kukkonen & Leonard, 2014; Leonard & Kukkonen, 2014), and thus it will be only briefly summarized here. Orexin receptor signaling appears to be very versatile. Both receptors have been reported to be able to couple to at least three heterotrimeric G-protein families, G_s, $G_{i/o}$ and G_q, and β-arrestin (Evans et al., 2001; Karteris et al., 2001; Holmqvist et al., 2005; Magga et al., 2006; Karteris et al., 2005; Milasta et al., 2005; Jaeger, Seeber, Eidne, & Pfleger, 2014; Kukkonen, 2016a; reviewed in Kukkonen, 2013; Kukkonen & Leonard, 2014; Leonard & Kukkonen, 2014). In addition, interactions via dynein light chain Tctex-types 1 and 3 and, upon receptor phosphorylation, with protein phosphatase SHP-2, have been suggested for OX_1 (Duguay et al., 2011; El Firar et al., 2009; Voisin, El Firar, Rouyer-Fessard, Gratio, & Laburthe, 2008). Orexin receptors also form complexes with themselves and with other GPCRs, which may modulate their signaling properties and pharmacology (see section "Orexin Receptor Complexes with Other Proteins") (Kukkonen, 2017). Nevertheless, orexin receptors are usually referred as G_q coupled—and most responses in recombinant CHO-K1 cells are mediated by G_q (Kukkonen, 2016a)—but the other G-protein (and other mediator) pathways should not be disregarded. It seems that different primary signal transducers are utilized in different cell types, but there is little information about this, especially from primary cells.

In the CNS, orexin responses have typically been measured with respect to neuronal excitation. Orexin receptors are neuroexcitatory via at least activation of postsynaptic nonselective cation fluxes and inhibition of K^+ channels, but some studies also report presynaptic effects, e.g., less well characterized enhancement of transmitter release as well as endocannabinoid generation, which may not be very surprising for a strongly G_q-coupled receptor (reviewed in Kukkonen, 2013; Kukkonen & Leonard, 2014; Leonard & Kukkonen, 2014; Kukkonen, 2017). In a somewhat longer timescale, synaptic plasticity is modulated by orexins (reviewed in Baimel & Borgland, 2012; Leonard & Kukkonen, 2014).

Another interesting property of the orexin receptors is their ability to induce programmed cell death in nonneuronal cell lines—neurons have not been tested—as indicated under *Cancer*.

OREXIN RECEPTOR COMPLEXES WITH OTHER PROTEINS

GPCRs are generally assumed to exist in di- or oligomeric complexes with themselves (homomeric) or other GPCRs (heteromeric) (Gaitonde & Gonzalez-Maeso, 2017; Gomes et al., 2016; Maggio, Innamorati, & Parenti, 2007; Milligan, 2009). Whether the number of components in these is two or more is usually not known—and the methods available do not easily allow distinction—and thus we here just call these dimers. The physiological significance of the GPCR dimerization/oligomerization is unknown for most GPCRs, though there are notable exceptions, in which the complexing, especially heteromerization, strongly affects maturation/trafficking, signaling, or pharmacology (Gomes et al., 2016; Milligan, 2009).

Orexin receptors appear to readily form complexes with themselves (homo- and heteromeric) and with other GPCRs including CB_1 cannabinoid, κ opioid, CRF_1 corticotropin-releasing factor and GPR103 receptors, though most of the evidence is for OX_1 only (Ward, Pediani, & Milligan, 2011; Xu, Ward, Pediani, & Milligan, 2011; Jäntti, Mandrika, & Kukkonen, 2014; Chen et al., 2015; Davies et al., 2015; Navarro et al., 2015; Robinson & McDonald, 2015; reviewed Kukkonen, 2017). The evidence comes from heterologous expression systems, and has not been physiologically verified, except for CRF_1 (Navarro et al., 2015). In addition, there have been no reports of some GPCR not interacting with orexin receptors in heterologous systems. Thus it is difficult to judge whether the di-/oligomerization of orexin receptors is a heterologous expression artifact or a phenomenon of physiological and pharmacological interest—except for the OX_1-CFR_1-σ_1 complex (Navarro et al., 2015).

Complexing of OX_1 with CFR_1 and σ_1 receptors in ventral tegmental area has a profound effect on the physiological readout, dopamine release (Navarro et al., 2015), but the molecular determinants of this downstream from the receptor complex are not known. There also is some evidence that OX_1 receptor dimerization with κ opioid receptor affects the primary signaling of OX_1 (Chen et al., 2015). However, we should be careful with the conclusions as GPCRs also show functional interaction independent of dimerization, i.e., mediated by the signal transduction pathways. Such is indicated for OX_1 and κ opioid receptors (Robinson & McDonald, 2015) as well as for OX_1 and CB_1 receptors (Haj-Dahmane & Shen, 2005; Ho et al., 2011; Jäntti et al., 2013). No pharmacological differences have been reported for any di-/oligomeric orexin receptor complexes. However, it must be pointed out that the signaling profiles or pharmacology has not been investigated for most of the orexin receptor complexes. An interesting approach to selectively target dimeric receptors are so-called bivalent ligands, i.e., ligands composed of two spacer-separated ends, each with a ligand with selectivity for one of the receptors. In this concept, each binding site is supposed to be intact or nearly so, so with careful adjustment of the ligand affinities or availability, selectivity can be obtained for the receptor dimer, which should display higher total affinity than either receptor alone. The ends can be designed to target the same (homobivalent ligands) or different (heterobivalent ligands) receptors, and either ligand can be an agonist or an antagonist (Gomes et al., 2016). If the active sites of the receptors are supposed to swap domains affecting the binding site (a hybrid binding site), then even a single, small molecular ligand might have affinity for the heterodimer but not for any monomeric receptor. A single study with bivalent ligands has been published for orexin receptors with no clear results (Perrey, Gilmour, Thomas, & Zhang, 2014).

Orexin receptors, like other GPCRs, also interact with other nonmembrane and membrane proteins than GPCRs. For orexin receptor, the obvious examples are the signal transducers, such as heterotrimeric G-proteins and β-arrestin (and potential other ones described under "Pharmacology of Orexin Receptor Signaling"). Such interactions could in principle also affect similar receptor properties as dimerization/oligomerization with another GPCR. The most often seen phenomenon is called biased signaling (Kenakin, 2011). In biased signaling, a receptor bound to a specific agonist is biased toward activation of a specific signaling pathway, whereas another agonist prefers another pathway. This is usually demonstrated for different G-proteins or β-arrestin (Kenakin, 2011). It is unclear, in which degree this comes from preformed complexes of receptors with signal transducers and in which degree the agonist-receptor complex, upon formation, interacts preferentially with specific signal transducers.

We have reported indications of biased signaling of orexin receptors with respect to different signal transduction pathways in CHO-K1 cells

(Kukkonen, 2013; Leonard & Kukkonen, 2014; Rinne et al., 2018) and across cell lines with respect to the same signal cascades (Putula et al., 2011). Such findings are by no means surprising, since for chemically sufficiently different agonists for receptors with ability to couple to several signal transducers, biased agonism is predicted. For such receptors, such as orexin receptors, care should therefore be taken in drug discovery. The primary signal transducer for orexin receptor signaling in the brain has not been identified but has been assumed to be G_q—though based on little evidence (see section "Pharmacology of Orexin Receptor Signaling"). If this is the case, then the often utilized screening method, Ca^{2+} elevation assay, may be suitable (Kukkonen, 2016a). We might, instead of or additionally to the brain functions, target cancer in our orexin receptor agonist discovery. It has been suggested that G_q also is behind this process (Voisin et al., 2008), but alternative cell death pathways for orexin receptors have been indicated as well (Kukkonen & Leonard, 2014) so the situation is not clear. It is thus strongly recommended that the orexin receptor coupling to—at least—the primary signal transducer for each physiological response of interest is verified and the screening assay designed according to this. For the $G_{q/11}$ coupling, there is a promising, recently validated inhibitor FR900359/UBO-QIC (Kukkonen, 2016b; Schrage et al., 2015).

In summary, it is likely that orexin receptors have an inherent ability to form heteromeric complexes with other GPCRs, and it is likely that such complexes also have an effect on the orexin receptor properties. Preferential signaling of orexin receptors has been shown to vary between tissues (Kukkonen, 2017; Kukkonen & Leonard, 2014). Such has previously been noted for GPCRs capable of coupling to several primary signal transducers, but how this is accomplished on the molecular level is not known. While there are differences in GPCR signal transducer—such as G-protein—expression in different cell types, this is unlikely to fully explain the coupling differences. It is possible that other proteins determine the receptor–primary signal transducer coupling. One idea is that the primary role of GPCR dimerization/oligomerization is to determine the G-protein coupling preference of the receptor complex.

CONCLUSIONS

As briefly inferred previously and in more detail in other chapters of this book, orexin receptors have taken a central stance in many physiological processes within the CNS. Most central of these is the regulation of sleep and wakefulness but there are also many other ones shown or speculated.

Yet more have been speculated for the periphery of the body. Accordingly, orexin receptors have attracted attention as drug discovery targets, and multiple antagonists have been and are being developed as sleep aids. This a very potential pharmacotherapeutic field, and it will be interesting to see the long-term efficacy and safety of, e.g., the single compound currently in therapeutic use, suvorexant. If orexins indeed are so intimately involved in many processes in the brain, we might expect significant side effects. On the other hand, in some of the physiological functions interesting for orexins, such as metabolic regulation, rodents are much different from human, and also the animal tests for some CNS processes such as depression and anxiety are quite rudimentary; thus not all the presumed functions have significance for human. For experimental and diagnostic purposes, I wish that some of the antagonists make their way to radiolabeled compounds for use in receptor binding and PET.

From both the experimental and therapeutic points of view, the lack of orexin receptor agonists is unlucky, and I hope this is mended in near future; the first series of promising compounds has been published. Orexin receptors are promiscuous GPCRs and thus for agonist activity in vivo the proper bias toward the signal pathway of interest is essential, and may in optimal case also reduce side effects. The current state of knowledge is that most responses are mediated by the G_q-family G-proteins, but since this is not based on much factual evidence, there is room (and need!) for proper studies with state-of-the-art tools. If orexin receptors indeed willingly make heteromeric complexes with other GPCRs, then bivalent/multifunctional ligands may be an interesting approach to increase the effect and reduce the side effects of orexinergic pharmacotherapy. However, our knowledge of the molecular details of the action of such ligands is limited, and the current ways of generating such ligands do not produce drugs with optimal classical ADME properties.

ACKNOWLEDGMENTS

This work was supported by the Magnus Ehrnrooth Foundation and the Liv & Hälsa Foundation.

REFERENCES

Abbas, M. G., Shoji, H., Soya, S., Hondo, M., Miyakawa, T., & Sakurai, T. (2015). Comprehensive behavioral analysis of male Ox1r (-/-) mice showed implication of orexin receptor-1 in mood, anxiety, and social behavior. *Frontiers in Behavioral Neuroscience*, *9*, 324. https://doi.org/10.3389/fnbeh.2015.00324.

Akbari, N., Salmani, M. E., Goudarzvand, M., LashkarBoluki, T., Goudarzi, I., & Abrari, K. (2014). Unilateral hypothalamus inactivation prevents PTZ kindling development through hippocampal orexin receptor 1 modulation. *Basic and Clinical Neuroscience, 5* (1), 66–73.

Al-Barazanji, K. A., Wilson, S., Baker, J., Jessop, D. S., & Harbuz, M. S. (2001). Central orexin-A activates hypothalamic-pituitary-adrenal axis and stimulates hypothalamic corticotropin releasing factor and arginine vasopressin neurones in conscious rats. *Journal of Neuroendocrinology, 13*(5), 421–424.

Alexandre, D., Hautot, C., Mehio, M., Jeandel, L., Courel, M., Voisin, T., et al. (2014). The orexin type 1 receptor is overexpressed in advanced prostate cancer with a neuroendocrine differentiation, and mediates apoptosis. *European Journal of Cancer, 50*(12), 2126–2133. https://doi.org/10.1016/j.ejca.2014.05.008.

Alvarez-Crespo, M., Martinez-Sanchez, N., Ruiz-Pino, F., Garcia-Lavandeira, M., Alvarez, C. V., Tena-Sempere, M., et al. (2013). The orexigenic effect of orexin-A revisited: Dependence of an intact growth hormone axis. *Endocrinology, 154*(10), 3589–3598. https://doi.org/10.1210/en.2013-1251.

Ammoun, S., Holmqvist, T., Shariatmadari, R., Oonk, H. B., Detheux, M., Parmentier, M., et al. (2003). Distinct recognition of OX1 and OX2 receptors by orexin peptides. *The Journal of Pharmacology and Experimental Therapeutics, 305*(2), 507–514. https://doi.org/10.1124/jpet.102.048025.

Ammoun, S., Johansson, L., Åkerman, K. E. O., Korhonen, L., Lindholm, D., & Kukkonen, J. P. (2003). *Regulation of MAP/SAP-kinase cascades by orexin receptors. Paper presented at the neuroscience 2003, New Orleans, LA.*

Ammoun, S., Lindholm, D., Wootz, H., Åkerman, K. E., & Kukkonen, J. P. (2006). G-protein-coupled OX1 orexin/hcrtr-1 hypocretin receptors induce caspase-dependent and -independent cell death through p38 mitogen-/stress-activated protein kinase. *The Journal of Biological Chemistry, 281*(2), 834–842. https://doi.org/10.1074/jbc.M508603200.

Aston-Jones, G., Smith, R. J., Sartor, G. C., Moorman, D. E., Massi, L., Tahsili-Fahadan, P., et al. (2010). Lateral hypothalamic orexin/hypocretin neurons: A role in reward-seeking and addiction. *Brain Research, 1314*, 74–90. https://doi.org/10.1016/j.brainres.2009.09.106.

Avolio, E., Biasone, A., Mele, M., & Alo, R. (2014). Distinct anxiogenic/anxiolytic effects exerted by the hamster lateral amygdalar nucleus injected with ORX-A or ORX-B in the presence of a GABAergic agonist. *Neuroreport, 25*(12), 932–937. https://doi.org/10.1097/WNR.0000000000000213.

Baimel, C., Bartlett, S. E., Chiou, L. C., Lawrence, A. J., Muschamp, J. W., Patkar, O., et al. (2015). Orexin/hypocretin role in reward: Implications for opioid and other addictions. *British Journal of Pharmacology, 172*(2), 334–348. https://doi.org/10.1111/bph.12639.

Baimel, C., & Borgland, S. L. (2012). Hypocretin modulation of drug-induced synaptic plasticity. *Progress in Brain Research, 198*, 123–131. https://doi.org/10.1016/B978-0-444-59489-1.00008-2.

Baimel, C., & Borgland, S. L. (2017). Hypocretin/orexin and plastic adaptations associated with drug abuse. *Current Topics in Behavioral Neurosciences, 33*, 283–304. https://doi.org/10.1007/7854_2016_44.

Baimel, C., Lau, B. K., Qiao, M., & Borgland, S. L. (2017). Projection-target-defined effects of orexin and dynorphin on VTA dopamine neurons. *Cell Reports, 18*(6), 1346–1355. https://doi.org/10.1016/j.celrep.2017.01.030.

Berthoud, H. R., Patterson, L. M., Sutton, G. M., Morrison, C., & Zheng, H. (2005). Orexin inputs to caudal raphe neurons involved in thermal, cardiovascular, and gastrointestinal regulation. *Histochemistry and Cell Biology, 123*(2), 147–156. https://doi.org/10.1007/s00418-005-0761-x.

Bingham, S., Davey, P. T., Babbs, A. J., Irving, E. A., Sammons, M. J., Wyles, M., et al. (2001). Orexin-A, an hypothalamic peptide with analgesic properties. *Pain*, *92*(1-2), 81–90. https://doi.org/10.1016/S0304-3959(00)00470-X.

Bonaventure, P., Dugovic, C., Shireman, B., Preville, C., Yun, S., Lord, B., et al. (2017). Evaluation of JNJ-54717793 a novel brain penetrant selective orexin 1 receptor antagonist in two rat models of panic attack provocation. *Frontiers in Pharmacology*, *8*, 357. https://doi.org/10.3389/fphar.2017.00357.

Borgland, S. L., Storm, E., & Bonci, A. (2008). Orexin B/hypocretin 2 increases glutamatergic transmission to ventral tegmental area neurons. *The European Journal of Neuroscience*, *28*(8), 1545–1556. https://doi.org/10.1111/j.1460-9568.2008.06397.x.

Borgland, S. L., Taha, S. A., Sarti, F., Fields, H. L., & Bonci, A. (2006). Orexin A in the VTA is critical for the induction of synaptic plasticity and behavioral sensitization to cocaine. *Neuron*, *49*(4), 589–601.

Boutrel, B., Kenny, P. J., Specio, S. E., Martin-Fardon, R., Markou, A., Koob, G. F., et al. (2005). Role for hypocretin in mediating stress-induced reinstatement of cocaine-seeking behavior. *Proceedings of the National Academy of Sciences of the United States of America*, *102*(52), 19168–19173. https://doi.org/10.1073/pnas.0507480102.

Brunton, P. J., & Russell, J. A. (2003). Hypothalamic-pituitary-adrenal responses to centrally administered orexin-A are suppressed in pregnant rats. *Journal of Neuroendocrinology*, *15* (7), 633–637.

Bunney, P. E., Zink, A. N., Holm, A. A., Billington, C. J., & Kotz, C. M. (2017). Orexin activation counteracts decreases in nonexercise activity thermogenesis (NEAT) caused by high-fat diet. *Physiology & Behavior*, *176*, 139–148. https://doi.org/10.1016/j.physbeh.2017.03.040.

Callander, G. E., Olorunda, M., Monna, D., Schuepbach, E., Langenegger, D., Betschart, C., et al. (2013). Kinetic properties of "dual" orexin receptor antagonists at OX1R and OX2R orexin receptors. *Frontiers in Neuroscience*, 7, 230. https://doi.org/10.3389/fnins.2013.00230.

Cano, M., Grima, P. M., & Benet, A. P. (2014). *US 2010/0150840 A1*.

Cason, A. M., Smith, R. J., Tahsili-Fahadan, P., Moorman, D. E., Sartor, G. C., & Aston-Jones, G. (2010). Role of orexin/hypocretin in reward-seeking and addiction: Implications for obesity. *Physiology & Behavior*, *100*(5), 419–428. https://doi.org/10.1016/j.physbeh.2010.03.009.

Chemelli, R. M., Willie, J. T., Sinton, C. M., Elmquist, J. K., Scammell, T., Lee, C., et al. (1999). Narcolepsy in orexin knockout mice: Molecular genetics of sleep regulation. *Cell*, *98*(4), 437–451. https://doi.org/10.1016/S0092-8674(00)81973-X.

Chen, C., & Xu, R. (2003). The in vitro regulation of growth hormone secretion by orexins. *Endocrine*, *22*(1), 57–66.

Chen, J., Zhang, R., Chen, X., Wang, C., Cai, X., Liu, H., et al. (2015). Heterodimerization of human orexin receptor 1 and kappa opioid receptor promotes protein kinase A/cAMP-response element binding protein signaling via a Galphas-mediated mechanism. *Cellular Signalling*, *27*(7), 1426–1438. https://doi.org/10.1016/j.cellsig.2015.03.027.

Coleman, P. J., & Renger, J. J. (2010). Orexin receptor antagonists: A review of promising compounds patented since 2006. *Expert Opinion on Therapeutic Patents*, *20*(3), 307–324. https://doi.org/10.1517/13543770903567085.

Dahmen, N., Bierbrauer, J., & Kasten, M. (2001). Increased prevalence of obesity in narcoleptic patients and relatives. *European Archives of Psychiatry and Clinical Neuroscience*, *251* (2), 85–89.

Dahmen, N., Tonn, P., Messroghli, L., Ghezel-Ahmadi, D., & Engel, A. (2009). Basal metabolic rate in narcoleptic patients. *Sleep*, *32*(7), 962–964.

Darker, J. G., Porter, R. A., Eggleston, D. S., Smart, D., Brough, S. J., Sabido-David, C., et al. (2001). Structure-activity analysis of truncated orexin-A analogues at the orexin-1 receptor. *Bioorganic & Medicinal Chemistry Letters*, *11*(5), 737–740.

Davies, J., Chen, J., Pink, R., Carter, D., Saunders, N., Sotiriadis, G., et al. (2015). Orexin receptors exert a neuroprotective effect in Alzheimer's disease (AD) via heterodimerization with GPR103. *Scientific Reports*, *5*, 12584. https://doi.org/10.1038/srep12584.

de Lecea, L. (2015). Optogenetic control of hypocretin (orexin) neurons and arousal circuits. *Current Topics in Behavioral Neurosciences*, *25*, 367–378. https://doi.org/10.1007/7854_2014_364.

de Lecea, L., Kilduff, T. S., Peyron, C., Gao, X., Foye, P. E., Danielson, P. E., et al. (1998). The hypocretins: Hypothalamus-specific peptides with neuroexcitatory activity. *Proceedings of the National Academy of Sciences of the United States of America*, *95*(1), 322–327.

Deadwyler, S. A., Porrino, L., Siegel, J. M., & Hampson, R. E. (2007). Systemic and nasal delivery of orexin-A (Hypocretin-1) reduces the effects of sleep deprivation on cognitive performance in nonhuman primates. *The Journal of Neuroscience*, *27*(52), 14239–14247. https://doi.org/10.1523/JNEUROSCI.3878-07.2007.

Duguay, D., Belanger-Nelson, E., Mongrain, V., Beben, A., Khatchadourian, A., & Cermakian, N. (2011). Dynein light chain Tctex-type 1 modulates orexin signaling through its interaction with orexin 1 receptor. *PLoS One*, *6*(10), e26430. https://doi.org/10.1371/journal.pone.0026430.

El Firar, A., Voisin, T., Rouyer-Fessard, C., Ostuni, M. A., Couvineau, A., & Laburthe, M. (2009). Discovery of a functional immunoreceptor tyrosine-based switch motif in a 7-transmembrane-spanning receptor: Role in the orexin receptor OX1R-driven apoptosis. *The FASEB Journal*, *23*(12), 4069–4080. https://doi.org/10.1096/fj.09-131367.

Erken, H. A., Erken, G., Genc, O., Kortunay, S., Sahiner, M., Turgut, G., et al. (2012). Orexins cause epileptic activity. *Peptides*, *37*(1), 161–164. https://doi.org/10.1016/j.peptides.2012.06.012.

Evans, N. A., Groarke, D. A., Warrack, J., Greenwood, C. J., Dodgson, K., Milligan, G., et al. (2001). Visualizing differences in ligand-induced beta-arrestin-GFP interactions and trafficking between three recently characterized G protein-coupled receptors. *Journal of Neurochemistry*, 77(2), 476–485.

Fadel, J., Bubser, M., & Deutch, A. Y. (2002). Differential activation of orexin neurons by antipsychotic drugs associated with weight gain. *The Journal of Neuroscience*, *22*(15), 6742–6746.

Flores, A., Valls-Comamala, V., Costa, G., Saravia, R., Maldonado, R., & Berrendero, F. (2014). The hypocretin/orexin system mediates the extinction of fear memories. *Neuropsychopharmacology*, *39*(12), 2732–2741. https://doi.org/10.1038/npp.2014.146.

Fronczek, R., Overeem, S., Reijntjes, R., Lammers, G. J., van Dijk, J. G., & Pijl, H. (2008). Increased heart rate variability but normal resting metabolic rate in hypocretin/orexin-deficient human narcolepsy. *Journal of Clinical Sleep Medicine*, *4*(3), 248–254.

Fujiki, N., Yoshida, Y., Ripley, B., Mignot, E., & Nishino, S. (2003). Effects of IV and ICV hypocretin-1 (orexin A) in hypocretin receptor-2 gene mutated narcoleptic dogs and IV hypocretin-1 replacement therapy in a hypocretin-ligand-deficient narcoleptic dog. *Sleep*, *26*(8), 953–959.

Fujiki, N., Yoshida, Y., Zhang, S., Sakurai, T., Yanagisawa, M., & Nishino, S. (2006). Sex difference in body weight gain and leptin signaling in hypocretin/orexin deficient mouse models. *Peptides*, *27*(9), 2326–2331. https://doi.org/10.1016/j.peptides.2006.03.011.

Funato, H., Tsai, A. L., Willie, J. T., Kisanuki, Y., Williams, S. C., Sakurai, T., et al. (2009). Enhanced orexin receptor-2 signaling prevents diet-induced obesity and improves leptin sensitivity. *Cell Metabolism*, *9*(1), 64–76. https://doi.org/10.1016/j.cmet.2008.10.010.

Gaitonde, S. A., & Gonzalez-Maeso, J. (2017). Contribution of heteromerization to G protein-coupled receptor function. *Current Opinion in Pharmacology*, *32*, 23–31. https://doi.org/10.1016/j.coph.2016.10.006.

Gao, M., Wang, M., & Zheng, Q. H. (2016). Synthesis of [(11)C]MK-1064 as a new PET radioligand for imaging of orexin-2 receptor. *Bioorganic & Medicinal Chemistry Letters*, *26* (15), 3694–3699. https://doi.org/10.1016/j.bmcl.2016.05.083.

Goforth, P. B., Leinninger, G. M., Patterson, C. M., Satin, L. S., & Myers, M. G., Jr. (2014). Leptin acts via lateral hypothalamic area neurotensin neurons to inhibit orexin neurons by multiple GABA-independent mechanisms. *The Journal of Neuroscience*, *34*(34), 11405–11415. https://doi.org/10.1523/JNEUROSCI.5167-13.2014.
Gomes, I., Ayoub, M. A., Fujita, W., Jaeger, W. C., Pfleger, K. D., & Devi, L. A. (2016). G Protein-coupled receptor heteromers. *Annual Review of Pharmacology and Toxicology*, *56*, 403–425. https://doi.org/10.1146/annurev-pharmtox-011613-135952.
Gomez, J. L., Bonaventura, J., Lesniak, W., Mathews, W. B., Sysa-Shah, P., Rodriguez, L. A., et al. (2017). Chemogenetics revealed: DREADD occupancy and activation via converted clozapine. *Science*, *357*(6350), 503–507. https://doi.org/10.1126/science.aan2475.
Gozzi, A., Turrini, G., Piccoli, L., Massagrande, M., Amantini, D., Antolini, M., et al. (2011). Functional magnetic resonance imaging reveals different neural substrates for the effects of orexin-1 and orexin-2 receptor antagonists. *PLoS One*, *6*(1), e16406. https://doi.org/10.1371/journal.pone.0016406.
Grafe, L. A., Eacret, D., Luz, S., Gotter, A. L., Renger, J. J., Winrow, C. J., et al. (2017). Orexin 2 receptor regulation of the hypothalamic-pituitary-adrenal (HPA) response to acute and repeated stress. *Neuroscience*, *348*, 313–323. https://doi.org/10.1016/j.neuroscience.2017.02.038.
Haj-Dahmane, S., & Shen, R. Y. (2005). The wake-promoting peptide orexin-B inhibits glutamatergic transmission to dorsal raphe nucleus serotonin neurons through retrograde endocannabinoid signaling. *The Journal of Neuroscience*, *25*(4), 896–905. https://doi.org/10.1523/JNEUROSCI.3258-04.2005.
Hara, J., Beuckmann, C. T., Nambu, T., Willie, J. T., Chemelli, R. M., Sinton, C. M., et al. (2001). Genetic ablation of orexin neurons in mice results in narcolepsy, hypophagia, and obesity. *Neuron*, *30*(2), 345–354.
Hara, J., Yanagisawa, M., & Sakurai, T. (2005). Difference in obesity phenotype between orexin-knockout mice and orexin neuron-deficient mice with same genetic background and environmental conditions. *Neuroscience Letters*, *380*(3), 239–242. https://doi.org/10.1016/j.neulet.2005.01.046.
Harris, G. C., Wimmer, M., & Aston-Jones, G. (2005). A role for lateral hypothalamic orexin neurons in reward seeking. *Nature*, *437*(7058), 556–559. https://doi.org/10.1038/nature04071.
Hasegawa, E., Yanagisawa, M., Sakurai, T., & Mieda, M. (2014). Orexin neurons suppress narcolepsy via 2 distinct efferent pathways. *Journal of Clinical Investigation*, *124*(2), 604–616. https://doi.org/10.1172/JCI71017.
Haynes, A. C., Jackson, B., Chapman, H., Tadayyon, M., Johns, A., Porter, R. A., et al. (2000). A selective orexin-1 receptor antagonist reduces food consumption in male and female rats. *Regulatory Peptides*, *96*(1-2), 45–51.
Ho, Y. C., Lee, H. J., Tung, L. W., Liao, Y. Y., Fu, S. Y., Teng, S. F., et al. (2011). Activation of orexin 1 receptors in the periaqueductal gray of male rats leads to antinociception via retrograde endocannabinoid (2-arachidonoylglycerol)-induced disinhibition. *The Journal of Neuroscience*, *31*(41), 14600–14610. https://doi.org/10.1523/JNEUROSCI.2671-11.2011.
Hoffmann, J., Supronsinchai, W., Akerman, S., Andreou, A. P., Winrow, C. J., Renger, J., et al. (2015). Evidence for orexinergic mechanisms in migraine. *Neurobiology of Disease*, *74*, 137–143. https://doi.org/10.1016/j.nbd.2014.10.022.
Holmqvist, T., Åkerman, K. E. O., & Kukkonen, J. P. (2002). Orexin signaling in recombinant neuron-like cells. *FEBS Letters*, *526*, 11–14. https://doi.org/10.1016/S0014-5793(02)03101-0.
Holmqvist, T., Johansson, L., Östman, M., Ammoun, S., Åkerman, K. E., & Kukkonen, J. P. (2005). OX1 orexin receptors couple to adenylyl cyclase regulation via multiple mechanisms. *The Journal of Biological Chemistry*, *280*(8), 6570–6579. https://doi.org/10.1074/jbc.M407397200.

Holohan, C., Van Schaeybroeck, S., Longley, D. B., & Johnston, P. G. (2013). Cancer drug resistance: An evolving paradigm. *Nature Reviews Cancer, 13*(10), 714–726. https://doi.org/10.1038/nrc3599.

Huang, S. C., Dai, Y. W., Lee, Y. H., Chiou, L. C., & Hwang, L. L. (2010). Orexins depolarize rostral ventrolateral medulla neurons and increase arterial pressure and heart rate in rats mainly via orexin 2 receptors. *The Journal of Pharmacology and Experimental Therapeutics, 334*(2), 522–529. https://doi.org/10.1124/jpet.110.167791.

Irukayama-Tomobe, Y., Ogawa, Y., Tominaga, H., Ishikawa, Y., Hosokawa, N., Ambai, S., et al. (2017). Nonpeptide orexin type-2 receptor agonist ameliorates narcolepsy-cataplexy symptoms in mouse models. *Proceedings of the National Academy of Sciences of the United States of America, 114*(22), 5731–5736. https://doi.org/10.1073/pnas.1700499114.

Jackson, K. L., Dampney, B. W., Moretti, J. L., Stevenson, E. R., Davern, P. J., Carrive, P., et al. (2016). Contribution of orexin to the neurogenic hypertension in BPH/2J mice. *Hypertension, 67*(5), 959–969. https://doi.org/10.1161/HYPERTENSIONAHA.115.07053.

Jaeger, W. C., Seeber, R. M., Eidne, K. A., & Pfleger, K. D. (2014). Molecular determinants of orexin receptor-arrestin-ubiquitin complex formation. *British Journal of Pharmacology, 171*(2), 364–374. https://doi.org/10.1111/bph.12481.

Jalewa, J., Wong-Lin, K., McGinnity, T. M., Prasad, G., & Holscher, C. (2014). Increased number of orexin/hypocretin neurons with high and prolonged external stress-induced depression. *Behavioural Brain Research, 272*, 196–204. https://doi.org/10.1016/j.bbr.2014.05.030.

James, M. H., Mahler, S. V., Moorman, D. E., & Aston-Jones, G. (2017). A decade of orexin/hypocretin and addiction: Where are we now? *Current Topics in Behavioral Neurosciences, 33*, 247–281. https://doi.org/10.1007/7854_2016_57.

Jann, M. W., Lam, Y. W., & Chang, W. H. (1994). Rapid formation of clozapine in guinea-pigs and man following clozapine-N-oxide administration. *Archives Internationales de Pharmacodynamie et de Thérapie, 328*(2), 243–250.

Jäntti, M. H., Mandrika, I., & Kukkonen, J. P. (2014). Human orexin/hypocretin receptors form constitutive homo- and heteromeric complexes with each other and with human CB1 cannabinoid receptors. *Biochemical and Biophysical Research Communications, 445*(2), 486–490. https://doi.org/10.1016/j.bbrc.2014.02.026.

Jäntti, M. H., Putula, J., Turunen, P. M., Näsman, J., Reijonen, S., & Kukkonen, J. P. (2013). Autocrine endocannabinoid signaling potentiates orexin receptor signaling upon CB1 cannabinoid-OX1 orexin receptor coexpression. *Molecular Pharmacology, 83*(3), 621–632. https://doi.org/10.1124/mol.112.080523.

John, J., Wu, M. F., & Siegel, J. M. (2000). Systemic administration of hypocretin-1 reduces cataplexy and normalizes sleep and waking durations in narcoleptic dogs. *Sleep Research Online, 3*(1), 23–28.

Johnson, P. L., Samuels, B. C., Fitz, S. D., Lightman, S. L., Lowry, C. A., & Shekhar, A. (2012). Activation of the orexin 1 receptor is a critical component of CO_2-mediated anxiety and hypertension but not bradycardia. *Neuropsychopharmacology, 37*(8), 1911–1922. https://doi.org/10.1038/npp.2012.38.

Johnson, P. L., Truitt, W., Fitz, S. D., Minick, P. E., Dietrich, A., Sanghani, S., et al. (2010). A key role for orexin in panic anxiety. *Nature Medicine, 16*(1), 111–115. https://doi.org/10.1038/nm.2075.

Kalogiannis, M., Hsu, E., Willie, J. T., Chemelli, R. M., Kisanuki, Y. Y., Yanagisawa, M., et al. (2011). Cholinergic modulation of narcoleptic attacks in double orexin receptor knockout mice. *PLoS One, 6*(4), e18697. https://doi.org/10.1371/journal.pone.0018697.

Karasawa, H., Yakabi, S., Wang, L., & Tache, Y. (2014). Orexin-1 receptor mediates the increased food and water intake induced by intracerebroventricular injection of the stable somatostatin pan-agonist, ODT8-SST in rats. *Neuroscience Letters*, *576*, 88–92. https://doi.org/10.1016/j.neulet.2014.05.063.

Karhu, L., Turku, A., & Xhaard, H. (2015). Modeling of the OX1R-orexin-A complex suggests two alternative binding modes. *BMC Structural Biology*, *15*, 9. https://doi.org/10.1186/s12900-015-0036-2.

Karteris, E., Machado, R. J., Chen, J., Zervou, S., Hillhouse, E. W., & Randeva, H. S. (2005). Food deprivation differentially modulates orexin receptor expression and signalling in the rat hypothalamus and adrenal cortex. *American Journal of Physiology Endocrinology and Metabolism*, *288*(6), E1089–E1100. https://doi.org/10.1152/ajpendo.00351.2004.

Karteris, E., Randeva, H. S., Grammatopoulos, D. K., Jaffe, R. B., & Hillhouse, E. W. (2001). Expression and coupling characteristics of the crh and orexin type 2 receptors in human fetal adrenals. *The Journal of Clinical Endocrinology and Metabolism*, *86*(9), 4512–4519.

Kastin, A. J., & Akerstrom, V. (1999). Orexin A but not orexin B rapidly enters brain from blood by simple diffusion. *The Journal of Pharmacology and Experimental Therapeutics*, *289*(1), 219–223.

Kenakin, T. (2011). Functional selectivity and biased receptor signaling. *The Journal of Pharmacology and Experimental Therapeutics*, *336*(2), 296–302. https://doi.org/10.1124/jpet.110.173948.

Khalil, R., & Fendt, M. (2017). Increased anxiety but normal fear and safety learning in orexin-deficient mice. *Behavioural Brain Research*, *320*, 210–218. https://doi.org/10.1016/j.bbr.2016.12.007.

Kim, M. K., Park, H. J., Kim, S. R., Choi, Y. K., Bae, S. K., & Bae, M. K. (2015). Involvement of heme oxygenase-1 in orexin-A-induced angiogenesis in vascular endothelial cells. *The Korean Journal of Physiology & Pharmacology*, *19*(4), 327–334. https://doi.org/10.4196/kjpp.2015.19.4.327.

Kok, S. W., Overeem, S., Visscher, T. L., Lammers, G. J., Seidell, J. C., Pijl, H., et al. (2003). Hypocretin deficiency in narcoleptic humans is associated with abdominal obesity. *Obesity Research*, *11*(9), 1147–1154.

Kok, S. W., Roelfsema, F., Overeem, S., Lammers, G. J., Strijers, R. L., Frolich, M., et al. (2002). Dynamics of the pituitary-adrenal ensemble in hypocretin-deficient narcoleptic humans: Blunted basal adrenocorticotropin release and evidence for normal time-keeping by the master pacemaker. *The Journal of Clinical Endocrinology and Metabolism*, *87*(11), 5085–5091.

Kordi Jaz, E., Moghimi, A., Fereidoni, M., Asadi, S., Shamsizadeh, A., & Roohbakhsh, A. (2017). SB-334867, an orexin receptor 1 antagonist, decreased seizure and anxiety in pentylenetetrazol-kindled rats. *Fundamental and Clinical Pharmacology*, *31*(2), 201–207. https://doi.org/10.1111/fcp.12249.

Korotkova, T. M., Sergeeva, O. A., Eriksson, K. S., Haas, H. L., & Brown, R. E. (2003). Excitation of ventral tegmental area dopaminergic and nondopaminergic neurons by orexins/hypocretins. *The Journal of Neuroscience*, *23*(1), 7–11.

Kortunay, S., Erken, H. A., Erken, G., Genc, O., Sahiner, M., Turgut, S., et al. (2012). Orexins increase penicillin-induced epileptic activity. *Peptides*, *34*(2), 419–422. https://doi.org/10.1016/j.peptides.2012.02.013.

Kotagal, S., Krahn, L. E., & Slocumb, N. (2004). A putative link between childhood narcolepsy and obesity. *Sleep Medicine*, *5*(2), 147–150. https://doi.org/10.1016/j.sleep.2003.10.006.

Kukkonen, J. P. (2012). Recent progress in orexin/hypocretin physiology and pharmacology. *Biomolecular Concepts*, *3*(5), 447–463. https://doi.org/10.1515/bmc-2012-0013.

Kukkonen, J. P. (2013). Physiology of the orexinergic/hypocretinergic system: A revisit in 2012. *American Journal of Physiology Cell Physiology*, *301*(1), C2–C32. https://doi.org/10.1152/ajpcell.00227.2012.

Kukkonen, J. P. (2016a). G-protein-dependency of orexin/hypocretin receptor signalling in recombinant Chinese hamster ovary cells. *Biochemical and Biophysical Research Communications*, *476*(4), 379–385. https://doi.org/10.1016/j.bbrc.2016.05.130.

Kukkonen, J. P. (2016b). G-protein inhibition profile of the reported Gq/11 inhibitor UBO-QIC. *Biochemical and Biophysical Research Communications*, *1*(1), 101–107. https://doi.org/10.1016/j.bbrc.2015.11.078.

Kukkonen, J. P. (2017). Orexin/hypocretin signaling. *Current Topics in Behavioral Neurosciences*, *33*, 17–50. https://doi.org/10.1007/7854_2016_49.

Kukkonen, J. P., Holmqvist, T., Ammoun, S., & Åkerman, K. E. (2002). Functions of the orexinergic/hypocretinergic system. *American Journal of Physiology Cell Physiology*, *283*(6), C1567–C1591. https://doi.org/10.1152/ajpcell.00055.2002.

Kukkonen, J. P., & Leonard, C. S. (2014). Orexin/hypocretin receptor signalling cascades. *British Journal of Pharmacology*, *171*(2), 294–313. https://doi.org/10.1111/bph.12324.

Kunii, K., Yamanaka, A., Nambu, T., Matsuzaki, I., Goto, K., & Sakurai, T. (1999). Orexins/hypocretins regulate drinking behaviour. *Brain Research*, *842*(1), 256–261. https://doi.org/10.1016/S0006-8993(99)01884-3.

Kuru, M., Ueta, Y., Serino, R., Nakazato, M., Yamamoto, Y., Shibuya, I., et al. (2000). Centrally administered orexin/hypocretin activates HPA axis in rats. *Neuroreport*, *11* (9), 1977–1980.

Langmead, C. J., Jerman, J. C., Brough, S. J., Scott, C., Porter, R. A., & Herdon, H. J. (2004). Characterisation of the binding of [3H]-SB-674042, a novel nonpeptide antagonist, to the human orexin-1 receptor. *British Journal of Pharmacology*, *141*(2), 340–346.

Lee, J., Reddy, M. M., & Kodadek, T. (2010). Discovery of an orexin receptor positive potentiator. *Chemical Science*. *1*(1). https://doi.org/10.1039/C0SC00197J.

Leonard, C. S., & Kukkonen, J. P. (2014). Orexin/hypocretin receptor signalling: A functional perspective. *British Journal of Pharmacology*, *171*(2), 294–313. https://doi.org/10.1111/bph.12296.

Li, A., Hindmarch, C. C., Nattie, E. E., & Paton, J. F. (2013). Antagonism of orexin receptors significantly lowers blood pressure in spontaneously hypertensive rats. *The Journal of Physiology*, *591*(Pt 17), 4237–4248. https://doi.org/10.1113/jphysiol.2013.256271.

Li, Y., Gao, X. B., Sakurai, T., & van den Pol, A. N. (2002). Hypocretin/orexin excites hypocretin neurons via a local glutamate neuron-A potential mechanism for orchestrating the hypothalamic arousal system. *Neuron*, *36*(6), 1169–1181.

Li, Y., Li, S., Wei, C., Wang, H., Sui, N., & Kirouac, G. J. (2010). Orexins in the paraventricular nucleus of the thalamus mediate anxiety-like responses in rats. *Psychopharmacology*, *212*(2), 251–265. https://doi.org/10.1007/s00213-010-1948-y.

Lin, L., Faraco, J., Li, R., Kadotani, H., Rogers, W., Lin, X., et al. (1999). The sleep disorder canine narcolepsy is caused by a mutation in the hypocretin (orexin) receptor 2 gene. *Cell*, *98*(3), 365–376. https://doi.org/10.1016/S0092-8674(00)81965-0.

Lin, Y., Matsumura, K., Tsuchihashi, T., Abe, I., & Iida, M. (2002). Chronic central infusion of orexin-A increases arterial pressure in rats. *Brain Research Bulletin*, *57*(5), 619–622. https://doi.org/10.1016/S0361-9230(01)00756-0.

Liu, F., Majo, V. J., Prabhakaran, J., Castrillion, J., Mann, J. J., Martinez, D., et al. (2012). Radiosynthesis of [11C]BBAC and [11C]BBPC as potential PET tracers for orexin2 receptors. *Bioorganic & Medicinal Chemistry Letters*, *22*(6), 2172–2174. https://doi.org/10.1016/j.bmcl.2012.01.114.

Lubkin, M., & Stricker-Krongrad, A. (1998). Independent feeding and metabolic actions of orexins in mice. *Biochemical and Biophysical Research Communications*, *253*(2), 241–245.

Lungwitz, E. A., Molosh, A., Johnson, P. L., Harvey, B. P., Dirks, R. C., Dietrich, A., et al. (2012). Orexin-A induces anxiety-like behavior through interactions with glutamatergic receptors in the bed nucleus of the stria terminalis of rats. *Physiology & Behavior*, *107*(5), 726–732. https://doi.org/10.1016/j.physbeh.2012.05.019.

Machado, B. H., Bonagamba, L. G., Dun, S. L., Kwok, E. H., & Dun, N. J. (2002). Pressor response to microinjection of orexin/hypocretin into rostral ventrolateral medulla of awake rats. *Regulatory Peptides*, *104*(1-3), 75–81.

MacLaren, D. A., Browne, R. W., Shaw, J. K., Krishnan Radhakrishnan, S., Khare, P., Espana, R. A., et al. (2016). Clozapine N-oxide administration produces behavioral effects in long-evans rats: Implications for designing DREADD experiments. *eNeuro*. *3*(5). https://doi.org/10.1523/ENEURO.0219-16.2016.

Magga, J., Bart, G., Oker-Blom, C., Kukkonen, J. P., Åkerman, K. E., & Näsman, J. (2006). Agonist potency differentiates G protein activation and Ca2+ signalling by the orexin receptor type 1. *Biochemical Pharmacology*, *71*(6), 827–836. https://doi.org/10.1016/j.bcp.2005.12.021.

Maggio, R., Innamorati, G., & Parenti, M. (2007). G protein-coupled receptor oligomerization provides the framework for signal discrimination. *Journal of Neurochemistry*, *103* (5), 1741–1752. https://doi.org/10.1111/j.1471-4159.2007.04896.x.

Malendowicz, L. K., Hochol, A., Ziolkowska, A., Nowak, M., Gottardo, L., & Nussdorfer, G. G. (2001). Prolonged orexin administration stimulates steroid-hormone secretion, acting directly on the rat adrenal gland. *International Journal of Molecular Medicine*, 7(4), 401–404.

Malendowicz, L. K., Tortorella, C., & Nussdorfer, G. G. (1999). Orexins stimulate corticosterone secretion of rat adrenocortical cells, through the activation of the adenylate cyclase-dependent signaling cascade. *The Journal of Steroid Biochemistry and Molecular Biology*, *70*(4-6), 185–188.

Malherbe, P., Borroni, E., Gobbi, L., Knust, H., Nettekoven, M., Pinard, E., et al. (2009). Biochemical and behavioural characterization of EMPA, a novel high-affinity, selective antagonist for the OX(2) receptor. *British Journal of Pharmacology*, *156*(8), 1326–1341. https://doi.org/10.1111/j.1476-5381.2009.00127.x.

Malherbe, P., Borroni, E., Pinard, E., Wettstein, J. G., & Knoflach, F. (2009). Biochemical and electrophysiological characterization of almorexant, a dual orexin 1 receptor (OX1)/orexin 2 receptor (OX2) antagonist: Comparison with selective OX1 and OX2 antagonists. *Molecular Pharmacology*, *76*(3), 618–631. https://doi.org/10.1124/mol.109.055152.

Malherbe, P., Roche, O., Marcuz, A., Kratzeisen, C., Wettstein, J. G., & Bissantz, C. (2010). Mapping the binding pocket of dual antagonist almorexant to human orexin 1 and orexin 2 receptors: Comparison with the selective OX1 (SB-674042) and OX2 (EMPA) antagonists. *Molecular Pharmacology*, *78*(1), 81–93. https://doi.org/10.1124/mol.110.064584.

Matsuki, T., Nomiyama, M., Takahira, H., Hirashima, N., Kunita, S., Takahashi, S., et al. (2009). Selective loss of GABA(B) receptors in orexin-producing neurons results in disrupted sleep/wakefulness architecture. *Proceedings of the National Academy of Sciences of the United States of America*, *106*(11), 4459–4464. https://doi.org/10.1073/pnas.0811126106.

Mazzocchi, G., Malendowicz, L. K., Gottardo, L., Aragona, F., & Nussdorfer, G. G. (2001). Orexin A stimulates cortisol secretion from human adrenocortical cells through activation of the adenylate cyclase-dependent signaling cascade. *The Journal of Clinical Endocrinology and Metabolism*, *86*(2), 778–782.

McElhinny, C. J., Jr., Lewin, A. H., Mascarella, S. W., Runyon, S., Brieaddy, L., & Carroll, F. I. (2012). Hydrolytic instability of the important orexin 1 receptor antagonist SB-334867: Possible confounding effects on in vivo and in vitro studies. *Bioorganic & Medicinal Chemistry Letters*, *22*(21), 6661–6664. https://doi.org/10.1016/j.bmcl.2012.08.109.

Mieda, M., Hasegawa, E., Kisanuki, Y. Y., Sinton, C. M., Yanagisawa, M., & Sakurai, T. (2011). Differential roles of orexin receptor-1 and -2 in the regulation of non-REM and REM sleep. *The Journal of Neuroscience*, *31*(17), 6518–6526. https://doi.org/10.1523/JNEUROSCI.6506-10.2011.

Mikrouli, E., Wortwein, G., Soylu, R., Mathe, A. A., & Petersen, S. (2011). Increased numbers of orexin/hypocretin neurons in a genetic rat depression model. *Neuropeptides*, *45* (6), 401–406. https://doi.org/10.1016/j.npep.2011.07.010.

Milasta, S., Evans, N. A., Ormiston, L., Wilson, S., Lefkowitz, R. J., & Milligan, G. (2005). The sustainability of interactions between the orexin-1 receptor and beta-arrestin-2 is defined by a single C-terminal cluster of hydroxy amino acids and modulates the kinetics of ERK MAPK regulation. *Biochemical Journal*, *387*(Pt 3), 573–584. https://doi.org/10.1042/BJ20041745.

Milella, M. S., Passarelli, F., De Carolis, L., Schepisi, C., Nativio, P., Scaccianoce, S., et al. (2010). Opposite roles of dopamine and orexin in quinpirole-induced excessive drinking: A rat model of psychotic polydipsia. *Psychopharmacology*, *211*(3), 355–366. https://doi.org/10.1007/s00213-010-1909-5.

Milligan, G. (2009). G protein-coupled receptor hetero-dimerization: Contribution to pharmacology and function. *British Journal of Pharmacology*, *158*(1), 5–14. https://doi.org/10.1111/j.1476-5381.2009.00169.x.

Nagahara, T., Saitoh, T., Kutsumura, N., Irukayama-Tomobe, Y., Ogawa, Y., Kuroda, D., et al. (2015). Design and synthesis of non-peptide, selective orexin receptor 2 agonists. *Journal of Medicinal Chemistry*, *58*(20), 7931–7937. https://doi.org/10.1021/acs.jmedchem.5b00988.

Nakamura, M., & Nagamine, T. (2017). Neuroendocrine, autonomic, and metabolic responses to an orexin antagonist, suvorexant, in psychiatric patients with insomnia. *Innovations in Clinical Neuroscience*, *14*(3–4), 30–37.

Nakamura, T., Uramura, K., Nambu, T., Yada, T., Goto, K., Yanagisawa, M., et al. (2000). Orexin-induced hyperlocomotion and stereotypy are mediated by the dopaminergic system. *Brain Research*, *873*(1), 181–187.

Navarro, G., Quiroz, C., Moreno-Delgado, D., Sierakowiak, A., McDowell, K., Moreno, E., et al. (2015). Orexin-corticotropin-releasing factor receptor heteromers in the ventral tegmental area as targets for cocaine. *The Journal of Neuroscience*, *35*(17), 6639–6653. https://doi.org/10.1523/JNEUROSCI.4364-14.2015.

Nishino, S. (2005). Hypocretin measurements in the CSF, and blood and brain tissue. Basic and clinical applications. In S. Nishino & T. Sakurai (Eds.), *Contemporary clinical neuroscience: The orexin/hypocretin system: Physiology and pathophysiology* (pp. 73–82). Totowa: Humana Press Inc.

Nowak, K. W., Mackowiak, P., Switonska, M. M., Fabis, M., & Malendowicz, L. K. (2000). Acute orexin effects on insulin secretion in the rat: In vivo and in vitro studies. *Life Sciences*, *66*(5), 449–454.

Nowak, K. W., Strowski, M. Z., Switonska, M. M., Kaczmarek, P., Singh, V., Fabis, M., et al. (2005). Evidence that orexins A and B stimulate insulin secretion from rat pancreatic islets via both receptor subtypes. *International Journal of Molecular Medicine*, *15*(6), 969–972.

Oi, N., Suzuki, M., Terauchi, T., Tokunaga, M., Nakatani, Y., Yamamoto, N., et al. (2013). Synthesis and evaluation of novel radioligands for positron emission tomography imaging of the orexin-2 receptor. *Journal of Medicinal Chemistry*, *56*(16), 6371–6385. https://doi.org/10.1021/jm400772t.

Oldfield, B. J., Giles, M. E., Watson, A., Anderson, C., Colvill, L. M., & McKinley, M. J. (2002). The neurochemical characterisation of hypothalamic pathways projecting polysynaptically to brown adipose tissue in the rat. *Neuroscience*, *110*(3), 515–526. https://doi.org/10.1016/S0306-4522(01)00555-3.

Ono, D., & Yamanaka, A. (2017). Hypothalamic regulation of the sleep/wake cycle. *Neuroscience Research*, *118*, 74–81. https://doi.org/10.1016/j.neures.2017.03.013.

Overeem, S., Kok, S. W., Lammers, G. J., Vein, A. A., Frolich, M., Meinders, A. E., et al. (2003). Somatotropic axis in hypocretin-deficient narcoleptic humans: Altered circadian distribution of GH-secretory events. *American Journal of Physiology Endocrinology and Metabolism*, *284*(3), E641–E647.

Park, J. H., Shim, H. M., Na, A. Y., Bae, J. H., Im, S. S., & Song, D. K. (2015). Orexin A regulates plasma insulin and leptin levels in a time-dependent manner following a glucose load in mice. *Diabetologia*, *58*(7), 1542–1550. https://doi.org/10.1007/s00125-015-3573-0.

Perez, M. V., Pavlovic, A., Shang, C., Wheeler, M. T., Miller, C. L., Liu, J., et al. (2015). Systems genomics identifies a key role for hypocretin/orexin receptor-2 in human heart failure. *Journal of the American College of Cardiology*, *66*(22), 2522–2533. https://doi.org/10.1016/j.jacc.2015.09.061.

Perrey, D. A., Gilmour, B. P., Thomas, B. F., & Zhang, Y. (2014). Toward the development of bivalent ligand probes of cannabinoid CB1 and orexin OX1 receptor heterodimers. *ACS Medicinal Chemistry Letters*, *5*(6), 634–638. https://doi.org/10.1021/ml4004759.

Peyron, C., Faraco, J., Rogers, W., Ripley, B., Overeem, S., Charnay, Y., et al. (2000). A mutation in a case of early onset narcolepsy and a generalized absence of hypocretin peptides in human narcoleptic brains. *Nature Medicine*, *6*(9), 991–997. https://doi.org/10.1038/79690.

Ponziani, V., Gennari, M., Pizza, F., Balsamo, A., Bernardi, F., & Plazzi, G. (2016). Growing up with type 1 narcolepsy: Its anthropometric and endocrine features. *Journal of Clinical Sleep Medicine*, *12*(12), 1649–1657. https://doi.org/10.5664/jcsm.6352.

Porter, R. A., Chan, W. N., Coulton, S., Johns, A., Hadley, M. S., Widdowson, K., et al. (2001). 1,3-Biarylureas as selective non-peptide antagonists of the orexin-1 receptor. *Bioorganic & Medicinal Chemistry Letters*, *11*(14), 1907–1910.

Putula, J., & Kukkonen, J. P. (2012). Mapping of the binding sites for the OX(1) orexin receptor antagonist, SB-334867, using orexin/hypocretin receptor chimaeras. *Neuroscience Letters*, *506*(1), 111–115. https://doi.org/10.1016/j.neulet.2011.10.061.

Putula, J., Turunen, P. M., Jäntti, M. H., Ekholm, M. E., & Kukkonen, J. P. (2011). Agonist ligand discrimination by the two orexin receptors depends on the expression system. *Neuroscience Letters*, *494*(1), 57–60. https://doi.org/10.1016/j.neulet.2011.02.055.

Ramanathan, L., & Siegel, J. M. (2014). Gender differences between hypocretin/orexin knockout and wild type mice: Age, body weight, body composition, metabolic markers, leptin and insulin resistance. *Journal of Neurochemistry*, *131*(5), 615–624. https://doi.org/10.1111/jnc.12840.

Rinne, M. K., Leino, T. O., Turku, A., Turunen, P. M., Steynen, Y., Xhaard, H., et al. (2018). Pharmacological characterization of the orexin/hypocretin receptor agonist Nag 26. *European Journal of Pharmacology*, *837*, 137–144. https://doi.org/10.1016/j.ejphar.2018.09.003.

Robinson, J. D., & McDonald, P. H. (2015). The orexin 1 receptor modulates kappa opioid receptor function via a JNK-dependent mechanism. *Cellular Signalling*, *27*(7), 1449–1456. https://doi.org/10.1016/j.cellsig.2015.03.026.

Roecker, A. J., Cox, C. D., & Coleman, P. J. (2016). Orexin receptor antagonists: New therapeutic agents for the treatment of insomnia. *Journal of Medicinal Chemistry*, *59*(2), 504–530. https://doi.org/10.1021/acs.jmedchem.5b00832.

Rouet-Benzineb, P., Rouyer-Fessard, C., Jarry, A., Avondo, V., Pouzet, C., Yanagisawa, M., et al. (2004). Orexins acting at native OX(1) receptor in colon cancer and neuroblastoma cells or at recombinant OX(1) receptor suppress cell growth by inducing apoptosis. *The Journal of Biological Chemistry*, *279*(44), 45875–45886. https://doi.org/10.1074/jbc.M404136200.

Sakurai, T., Amemiya, A., Ishii, M., Matsuzaki, I., Chemelli, R. M., Tanaka, H., et al. (1998). Orexins and orexin receptors: A family of hypothalamic neuropeptides and G protein-coupled receptors that regulate feeding behavior. *Cell*, *92*(4), 573–585. https://doi.org/10.1016/S0092-8674(00)80949-6.

Sakurai, T., Moriguchi, T., Furuya, K., Kajiwara, N., Nakamura, T., Yanagisawa, M., et al. (1999).. Structure and function of human prepro-orexin gene. *The Journal of Biological Chemistry*, *274*(25), 17771–17776.

Samson, W. K., Gosnell, B., Chang, J. K., Resch, Z. T., & Murphy, T. C. (1999). Cardiovascular regulatory actions of the hypocretins in brain. *Brain Research*, *831*(1-2), 248–253.

Samson, W. K., & Taylor, M. M. (2001). Hypocretin/orexin suppresses corticotroph responsiveness in vitro. *American Journal of Physiology Regulatory, Integrative and Comparative Physiology*, *281*(4), R1140–R1145.

Samson, W. K., Taylor, M. M., Follwell, M., & Ferguson, A. V. (2002). Orexin actions in hypothalamic paraventricular nucleus: Physiological consequences and cellular correlates. *Regulatory Peptides*, *104*(1-3), 97–103.

Sasaki, K., Suzuki, M., Mieda, M., Tsujino, N., Roth, B., & Sakurai, T. (2011). Pharmacogenetic modulation of orexin neurons alters sleep/wakefulness states in mice. *PLoS One*, *6*(5), e20360. https://doi.org/10.1371/journal.pone.0020360.

Schrage, R., Schmitz, A. L., Gaffal, E., Annala, S., Kehraus, S., Wenzel, D., et al. (2015). The experimental power of FR900359 to study Gq-regulated biological processes. *Nature Communications*, *6*, 10156. https://doi.org/10.1038/ncomms10156.

Schuld, A., Hebebrand, J., Geller, F., & Pollmacher, T. (2000). Increased body-mass index in patients with narcolepsy. *Lancet*, *355*(9211), 1274–1275.

Scott, M. M., Marcus, J. N., Pettersen, A., Birnbaum, S. G., Mochizuki, T., Scammell, T. E., et al. (2011). Hcrtr1 and 2 signaling differentially regulates depression-like behaviors. *Behavioural Brain Research*, *222*(2), 289–294. https://doi.org/10.1016/j.bbr.2011.02.044.

Sellayah, D., Bharaj, P., & Sikder, D. (2011). Orexin is required for brown adipose tissue development, differentiation, and function. *Cell Metabolism*, *14*(4), 478–490. https://doi.org/10.1016/j.cmet.2011.08.010.

Sellayah, D., & Sikder, D. (2012). Orexin receptor-1 mediates brown fat developmental differentiation. *Adipocytes*, *1*(1), 58–63. https://doi.org/10.4161/adip.18965.

Skrzypski, M., Kaczmarek, P., Le, T. T., Wojciechowicz, T., Pruszynska-Oszmalek, E., Szczepankiewicz, D., et al. (2012). Effects of orexin A on proliferation, survival, apoptosis and differentiation of 3T3-L1 preadipocytes into mature adipocytes. *FEBS Letters*, *586*(23), 4157–4164. https://doi.org/10.1016/j.febslet.2012.10.013.

Smart, D., Sabido-David, C., Brough, S. J., Jewitt, F., Johns, A., Porter, R. A., et al. (2001). SB-334867-A: The first selective orexin-1 receptor antagonist. *British Journal of Pharmacology*, *132*(6), 1179–1182.

Smith, P. M., Connolly, B. C., & Ferguson, A. V. (2002). Microinjection of orexin into the rat nucleus tractus solitarius causes increases in blood pressure. *Brain Research*, *950*(1–2), 261–267. https://doi.org/10.1016/S0006-8993(02)03048-2.

Smith, R. J., See, R. E., & Aston-Jones, G. (2009). Orexin/hypocretin signaling at the orexin 1 receptor regulates cue-elicited cocaine-seeking. *The European Journal of Neuroscience*, *30*(3), 493–503. https://doi.org/10.1111/j.1460-9568.2009.06844.x.

Socala, K., Szuster-Ciesielska, A., & Wlaz, P. (2016). SB 334867, a selective orexin receptor type 1 antagonist, elevates seizure threshold in mice. *Life Sciences*, *150*, 81–88. https://doi.org/10.1016/j.lfs.2016.02.075.

Soya, S., Takahashi, T. M., McHugh, T. J., Maejima, T., Herlitze, S., Abe, M., et al. (2017). Orexin modulates behavioral fear expression through the locus coeruleus. *Nature Communications*, *8*(1), 1606. https://doi.org/10.1038/s41467-017-01782-z.

Spinazzi, R., Ziolkowska, A., Neri, G., Nowak, M., Rebuffat, P., Nussdorfer, G. G., et al. (2005). Orexins modulate the growth of cultured rat adrenocortical cells, acting through type 1 and type 2 receptors coupled to the MAPK p42/p44- and p38-dependent cascades. *International Journal of Molecular Medicine*, *15*(5), 847–852.

Steiner, M. A., Sciarretta, C., Brisbare-Roch, C., Strasser, D. S., Studer, R., & Jenck, F. (2012). Examining the role of endogenous orexins in hypothalamus-pituitary-adrenal axis endocrine function using transient dual orexin receptor antagonism in the rat. *Psychoneuroendocrinology*. https://doi.org/10.1016/j.psyneuen.2012.07.016.

Suzuki, M., Beuckmann, C. T., Shikata, K., Ogura, H., & Sawai, T. (2005). Orexin-A (hypocretin-1) is possibly involved in generation of anxiety-like behavior. *Brain Research, 1044*(1), 116–121. https://doi.org/10.1016/j.brainres.2005.03.002.

Tabuchi, S., Tsunematsu, T., Black, S. W., Tominaga, M., Maruyama, M., Takagi, K., et al. (2014). Conditional ablation of orexin/hypocretin neurons: A new mouse model for the study of narcolepsy and orexin system function. *The Journal of Neuroscience, 34*(19), 6495–6509. https://doi.org/10.1523/JNEUROSCI.0073-14.2014.

Tang, J., Chen, J., Ramanjaneya, M., Punn, A., Conner, A. C., & Randeva, H. S. (2008). The signalling profile of recombinant human orexin-2 receptor. *Cellular Signalling, 20*(9), 1651–1661. https://doi.org/10.1016/j.cellsig.2008.05.010.

Thompson, M. D., Xhaard, H., Sakurai, T., Rainero, I., & Kukkonen, J. P. (2014). OX1 and OX2 orexin/hypocretin receptor pharmacogenetics. *Frontiers in Neuroscience, 8*, 57. https://doi.org/10.3389/fnins.2014.00057.

Tsuneki, H., Kon, K., Ito, H., Yamazaki, M., Takahara, S., Toyooka, N., et al. (2016). Timed inhibition of orexin system by suvorexant improved sleep and glucose metabolism in type 2 diabetic db/db mice. *Endocrinology, 157*(11), 4146–4157. https://doi.org/10.1210/en.2016-1404.

Tsuneki, H., Tokai, E., Nakamura, Y., Takahashi, K., Fujita, M., Asaoka, T., et al. (2015). Hypothalamic orexin prevents hepatic insulin resistance via daily bidirectional regulation of autonomic nervous system in mice. *Diabetes, 64*(2), 459–470. https://doi.org/10.2337/db14-0695.

Tupone, D., Madden, C. J., Cano, G., & Morrison, S. F. (2011). An orexinergic projection from perifornical hypothalamus to raphe pallidus increases rat brown adipose tissue thermogenesis. *The Journal of Neuroscience, 31*(44), 15944–15955. https://doi.org/10.1523/JNEUROSCI.3909-11.2011.

Turku, A., Borrel, A., Leino, T. O., Karhu, L., Kukkonen, J. P., & Xhaard, H. (2016). Pharmacophore model to discover OX1 and OX2 orexin receptor ligands. *Journal of Medicinal Chemistry, 59*(18), 8263–8275. https://doi.org/10.1021/acs.jmedchem.6b00333.

Turku, A., Rinne, M. K., Boije Af Gennas, G., Xhaard, H., Lindholm, D., & Kukkonen, J. P. (2017). Orexin receptor agonist Yan 7874 is a weak agonist of orexin/hypocretin receptors and shows orexin receptor-independent cytotoxicity. *PLoS One, 12*(6), e0178526. https://doi.org/10.1371/journal.pone.0178526.

Uramura, K., Funahashi, H., Muroya, S., Shioda, S., Takigawa, M., & Yada, T. (2001). Orexin-a activates phospholipase C- and protein kinase C-mediated Ca2+ signaling in dopamine neurons of the ventral tegmental area. *Neuroreport, 12*(9), 1885–1889.

Voisin, T., El Firar, A., Fasseu, M., Rouyer-Fessard, C., Descatoire, V., Walker, F., et al. (2011). Aberrant expression of OX1 receptors for orexins in colon cancers and liver metastases: An openable gate to apoptosis. *Cancer Research, 71*(9), 3341–3351. https://doi.org/10.1158/0008-5472.CAN-10-3473.

Voisin, T., El Firar, A., Rouyer-Fessard, C., Gratio, V., & Laburthe, M. (2008). A hallmark of immunoreceptor, the tyrosine-based inhibitory motif ITIM, is present in the G protein-coupled receptor OX1R for orexins and drives apoptosis: a novel mechanism. *The FASEB Journal, 22*(6), 1993–2002. https://doi.org/10.1096/fj.07-098723.

Voisin, T., Firar, A. E., Avondo, V., & Laburthe, M. (2006). Orexin-induced apoptosis: The key role of the seven-transmembrane domain orexin type 2 receptor. *Endocrinology, 147*(10), 4977–4984. https://doi.org/10.1210/en.2006-0201.

Wang, C., Moseley, C. K., Carlin, S. M., Wilson, C. M., Neelamegam, R., & Hooker, J. M. (2013). Radiosynthesis and evaluation of [11C]EMPA as a potential PET tracer for orexin 2 receptors. *Bioorganic & Medicinal Chemistry Letters, 23*(11), 3389–3392. https://doi.org/10.1016/j.bmcl.2013.03.079.

Wang, C., Pan, Y., Zhang, R., Bai, B., Chen, J., & Randeva, H. S. (2014). Heterodimerization of Mouse Orexin type 2 receptor variants and the effects on signal transduction. *Biochimica et Biophysica Acta, 1843*(3), 652–663. https://doi.org/10.1016/j.bbamcr.2013.12.010.

Wang, C., Wilson, C. M., Moseley, C. K., Carlin, S. M., Hsu, S., Arabasz, G., et al. (2013). Evaluation of potential PET imaging probes for the orexin 2 receptors. *Nuclear Medicine and Biology*, *40*(8), 1000–1005. https://doi.org/10.1016/j.nucmedbio.2013.07.001.

Ward, R. J., Pediani, J. D., & Milligan, G. (2011). Hetero-multimerization of the cannabinoid CB1 receptor and the orexin OX1 receptor generates a unique complex in which both protomers are regulated by orexin A. *The Journal of Biological Chemistry*, *286*(43), 37414–37428. https://doi.org/10.1074/jbc.M111.287649.

Willie, J. T., Chemelli, R. M., Sinton, C. M., Tokita, S., Williams, S. C., Kisanuki, Y. Y., et al. (2003). Distinct narcolepsy syndromes in Orexin receptor-2 and Orexin null mice: Molecular genetic dissection of Non-REM and REM sleep regulatory processes. *Neuron*, *38*(5), 715–730.

Willie, J. T., Chemelli, R. M., Sinton, C. M., & Yanagisawa, M. (2001). To eat or to sleep? Orexin in the regulation of feeding and wakefulness. *Annual Review of Neuroscience*, *24*, 429–458. https://doi.org/10.1146/annurev.neuro.24.1.429.

Winrow, C. J., Coleman, P., de Lecea, L., Kilduff, T., Kukkonen, J. P., Porter, R., et al. (2018a). *Orexin receptors: OX1 receptor.* Retrieved from http://www.guidetopharmacology.org/GRAC/ObjectDisplayForward?objectId=321. Accessed 23 January 2018.

Winrow, C. J., Coleman, P., de Lecea, L., Kilduff, T., Kukkonen, J. P., Porter, R., et al. (2018b). *Orexin receptors: OX2 receptor.* Retrieved from http://www.guidetopharmacology.org/GRAC/ObjectDisplayForward?objectId=322. Accessed 23 January 2018.

Winrow, C. J., Tanis, K. Q., Reiss, D. R., Rigby, A. M., Uslaner, J. M., Uebele, V. N., et al. (2010). Orexin receptor antagonism prevents transcriptional and behavioral plasticity resulting from stimulant exposure. *Neuropharmacology*, *58*(1), 185–194. https://doi.org/10.1016/j.neuropharm.2009.07.008.

Xu, T. R., Ward, R. J., Pediani, J. D., & Milligan, G. (2011). The orexin OX1 receptor exists predominantly as a homodimer in the basal state: Potential regulation of receptor organization by both agonist and antagonist ligands. *Biochemical Journal*, *439*(1), 171–183. https://doi.org/10.1042/BJ20110230.

Yanagisawa, M. (2010). USA Patent No. US 2010/0150840 A1.

Yin, J., Babaoglu, K., Brautigam, C. A., Clark, L., Shao, Z., Scheuermann, T. H., et al. (2016). Structure and ligand-binding mechanism of the human OX1 and OX2 orexin receptors. *Nature Structural & Molecular Biology*, *23*(4), 293–299. https://doi.org/10.1038/nsmb.3183.

Yin, J., Mobarec, J. C., Kolb, P., & Rosenbaum, D. M. (2015). Crystal structure of the human OX2 orexin receptor bound to the insomnia drug suvorexant. *Nature*, *519*(7542), 247–250. https://doi.org/10.1038/nature14035.

Zheng, H., Patterson, L. M., & Berthoud, H. R. (2005). Orexin-A projections to the caudal medulla and orexin-induced c-Fos expression, food intake, and autonomic function. *The Journal of Comparative Neurology*, *485*(2), 127–142. https://doi.org/10.1002/cne.20515.

Zink, A. N., Bunney, P. E., Holm, A. A., Billington, C. J., & Kotz, C. M. (2017). Neuromodulation of orexin neurons reduces diet-induced adiposity. *International Journal of Obesity*, *42*(4), 737–745. https://doi.org/10.1038/ijo.2017.276.

Zwirska-Korczala, K., Adamczyk-Sowa, M., Sowa, P., Pilc, K., Suchanek, R., Pierzchala, K., et al. (2007). Role of leptin, ghrelin, angiotensin II and orexins in 3T3 L1 preadipocyte cells proliferation and oxidative metabolism. *Journal of Physiology and Pharmacology*, *58*(Suppl 1), 53–64.

CHAPTER 3

Orexin Drives Energy Expenditure

Claudio Perez-Leighton*, **Jennifer A. Teske**[†], **Catherine M. Kotz**[‡,§]
*Department of Physiology, Faculty of Biological Sciences, Pontifical Catholic University of Chile, Chile
†Department of Nutritional Sciences, University of Arizona, Tucson, AZ, United States
‡Department of Integrative Biology and Physiology, University of Minnesota, Minneapolis, MN, United States
§GRECC, Minneapolis VA Health Care System, Minneapolis, MN, United States

INTRODUCTION

Orexin-Induced Spontaneous Physical Activity and Nonexercise Activity Thermogenesis

Energy expenditure (EE) in humans and mammals can be classified into several components, which include basal resting metabolism (BMR), thermogenesis, postprandial EE and EE due to physical activity. NEAT includes all forms of energy expenditure due to physical activity not associated with formal exercise, such as standing and fidgeting (Fruhbeck, 2005; Levine, 2002; Ravussin, 2005; Snitker, Tataranni, & Ravussin, 2001). NEAT contributes importantly to the variability to weight gain between humans (Levine, Eberhardt, & Jensen, 1999; Ravussin, Lillioja, Anderson, Christin, & Bogardus, 1986; Zurlo et al., 1992). In a seminal study showing the association between SPA and diet-induced obesity (DIO) sensitivity, participants were overfed 1000 kcal/day for 8 weeks. Under these conditions, not all individuals gained the same amount of fat mass. Instead, the amount of fat mass gain (which ranged from 0 to 5 kg) was negatively correlated with change in NEAT (Levine et al., 1999), which suggested that NEAT in humans is important to body weight regulation.

The phrase spontaneous physical activity (SPA) is used to describe all types of physical activity that contribute to NEAT. Thus SPA and NEAT are not interchangeable, but complementary concepts: NEAT refers to energy expenditure while SPA describes the types of physical activity that result in NEAT. In humans, SPA reflects an inherent drive for activity rather than goal-oriented activity and includes fidgeting, time spent standing, and ambulating (Kotz & Levine, 2005; Levine, Vander Weg, Hill, & Klesges, 2006; Garland et al., 2011). As NEAT protects against obesity in humans

The Orexin/Hypocretin System
https://doi.org/10.1016/B978-0-12-813751-2.00003-6

(Levine et al., 1999), so does SPA. For example, lean people spend larger amounts of time standing (a type of SPA) when compared to obese people (Levine et al., 2005); the time spent standing or sitting was unaffected by weight gain in the lean or weight loss in the obese, suggesting that standing time is an inherent trait. In rodents, SPA is measured as spontaneous ambulatory and rearing physical activity in an open field over a long period of time (i.e., 24 h) after adaptation to the new environment to avoid the confound effect of exploratory activity or novelty-driven anxiety (Garland et al., 2011; Kotz et al., 2006; Kotz, Teske, Levine, & Wang, 2002; Teske, Levine, Kuskowski, Levine, & Kotz, 2006). Therefore SPA represents a different type of physical activity than exercise, which in rodents is usually modeled by access to running wheel or treadmills (Garland et al., 2011). As in humans, higher SPA in rodents is associated with increased resistance to obesity (Novak, Kotz, & Levine, 2006; Teske et al., 2006; Teske, Billington, Kuskowski, & Kotz, 2012).

As described previously, SPA is physiologically relevant to human energy balance and is controlled by a distributed brain network that includes several neuropeptides and brain regions, among which the hypothalamic orexin neurons play a key role (Kotz, Teske, & Billington, 2008; Teske, Billington, & Kotz, 2008). These neural systems are yet to be fully described but may represent attractive therapeutic targets for obesity. In this review, we focus on evidence supporting a positive effect of orexins on SPA and NEAT.

ANIMAL MODELS OF OREXIN LOSS/GAIN AND SPA AND NEAT

Pharmacological Evidence

Early studies in rodents suggested that ICV injection of orexins increased physical activity (Hagan et al., 1999; Sweet, Levine, Billington, & Kotz, 1999). In one study, orexin-B-induced hyperactivity was blocked by SB-334867-A, an OX1 receptor antagonist, suggesting that OX1R mediates this response (Jones et al., 2001). Later studies showed that injection of orexin-A in different sites distributed across the brain can increase SPA, but not all the brain sites tested show the same sensitivity to orexin-A (Kotz et al., 2012; Novak et al., 2006; Perez-Leighton, Boland, et al., 2013; Teske et al., 2006). In rats, injection of orexin-A in the paraventricular hypothalamic nucleus (PVN), rostral lateral hypothalamus (LH), nucleus accumbens shell (NACsh) and substantia nigra pars compacta (SNpc), locus coeruleus (LC), dorsal raphe, and tuberomammillary nucleus increase SPA and/or feeding (Dube, Kalra, & Kalra, 1999; Kotz et al., 2002, 2008; Sweet et al., 1999; Thorpe & Kotz, 2005). Among these brain sites, PVN, rostral LH, TMN, and SNpc show equally high sensitivity to orexin-A-induced SPA in rats, though high activity/low activity (HA/LA) studies (described later in this review) suggest that the SNpc has lower sensitivity to orexin-A as compared to rostral LH (Kotz et al., 2008). Moreover, repeated injection of orexin-A in PVN and the rostral LH causes either weight loss or prevents obesity induced by high-fat diet intake (Kiwaki, Kotz, Wang, Lanningham-Foster, & Levine, 2004; Kotz et al., 2006; Novak & Levine, 2009; Perez-Leighton, Butterick-Peterson, et al., 2013; Perez-Leighton, Little, Grace, Billington, & Kotz, 2017). Together, this evidence supports the idea that orexin-A can drive SPA by its distributed action across different brain sites.

Pharmacological studies using orexin receptor antagonists support this role of orexin. Intraperitoneal injection of orexin receptor antagonists decreases SPA (Winrow et al., 2011, 2012). Moreover, orexin-A-stimulated SPA was blocked by injection of a selective orexin-1 receptor antagonist (SB-334867) in the rostral LH or SNpc (Teske, Billington, & Kotz, 2014) but not when injected into the NACsh (Thorpe & Kotz, 2005). A selective orexin-2 receptor antagonist (JNJ-10397049) in ventrolateral preoptic area (VLPO) (Mavanji et al., 2015) and a dual orexin receptor antagonist (TCS1102) in the VLPO (Coborn et al., 2017) also block orexin-A-induced SPA. Together, these data suggest that both OX1R and OX2R contribute to SPA and NEAT in a site-dependent manner.

Genetic Manipulations of Orexin Peptide and Receptors Expression

The earliest evidence from genetic manipulations indicating that orexin peptides are involved in NEAT was published in 2001. In this work, Hara and colleagues developed a mouse model that lacks orexin neurons (OX-AT3) (Hara et al., 2001). These mice express the ataxia3 gene under the control of the orexin promoter, leading to postnatal expression of the toxic protein ataxin3 in these neurons and subsequent neuronal death (Hara et al., 2001). These mice show both hypophagia and lower levels of SPA during the dark period, but the energetic consequences of lower activity are greater as they develop late onset obesity when fed a regular diet (Akiyama et al., 2004; Hara et al., 2001). Importantly, the development of obesity was observed concurrently with the loss of orexin neurons, suggesting that a role of the orexin neurons is to promote a net increase in energy expenditure. This hypothesis is supported by studies showing that chronic intrathecal orexin injections do not result in body weight gain, but rather weight loss (Kiwaki et al., 2004; Kotz et al., 2006; Novak & Levine, 2009; Perez-Leighton, Butterick-Peterson, et al., 2013; Perez-Leighton et al., 2017). An important caveat to this hypothesis was demonstrated in a study showing that mice deficient only in orexin peptides do not become obese (Hara, Yanagisawa, & Sakurai, 2005), suggesting that additional peptides expressed in orexin neurons contribute to promote NEAT. Yet other studies show that this effect may be diet dependent, since when fed a high-fat diet, mice deficient in orexin peptide expression gain more weight compared to wild-type mice (Sellayah, Bharaj, & Sikder, 2011). In addition, genetic background and sex moderate the orexin peptide functions. For example, the late spontaneous obesity observed in OX-AT3 mice, but not the higher weight gain on a high-fat diet, was dependent on genetic background (Hara et al., 2005). Both female OX-AT3 and OX −/− mice (from both C57BL/6 and mixed genetic background) gained more body weight on a standard diet compared to that in male mice (Fujiki et al., 2006). Together, this evidence supports the idea that the lack of orexin peptides in rodents leads to decreased energy expenditure by decreasing SPA and NEAT, and that this effect is influenced by diet, sex, and genetic background.

Data from studies with mice overexpressing the orexin peptides further support their role in promoting NEAT and a net energy expenditure. In these studies, mice with the prepro-orexin gene under the control of the β-actin/cytomegalovirus (CAG) promoter (CAG-OX), led to overexpression of the orexin peptides (Mieda et al., 2004) in the hypothalamus, although there was also evidence for ectopic orexin peptide expression (Funato et al., 2009). Male and female CAG-OX mice were more resistant to obesity induced by consumption of a high-fat diet compared to their wild-type littermates (Funato et al., 2009). The male CAG-OX mice showed higher EE and lower food intake compared to the wild-type mice, suggesting that in this model these two adaptations contributed to obesity resistance (Funato et al., 2009). When the CAG-OX mice were crossed to OX1R −/− and OX2R −/− mice, only OX1R −/− CAG-OX mice were resistant to diet-induced obesity (Funato et al., 2009), suggesting that the orexin 1 receptor may be the most influential in mediating obesity resistance. Yet, in the same study, chronic ICV infusion of an OX2R agonist promoted obesity resistance, indicating that activity of orexin 2 receptors may also confer obesity resistance. In another study supporting the role of OX2R in obesity resistance, repeated i.p. administration of YNT-185, a nonpeptide agonist of the OX2R in aged male mice lacking expression of the orexin peptides decreased weight compared to wild-type littermates, an effect that was reversed after administration of YNT-185 was terminated (Irukayama-Tomobe et al., 2017). This effect was not observed in mice deficient for both orexin receptors, thus confirming the specificity of the effect. These results suggest that activation of OX2R also mediates orexin defense against diet-induced obesity. Together, genetic evidence from these studies suggests that activation of the orexin system increases EE, which results in protection against development of obesity.

Direct Manipulation of Orexin Function: Chemogenetic and Optogenetic Studies

Activation of the hypothalamic orexin-producing neurons can increase SPA in male and female mice (Bunney, Zink, Holm, Billington, & Kotz, 2017; Inutsuka et al., 2014; Zink, Bunney, Holm, Billington, & Kotz, 2018). High-fat diet intake decreases SPA and NEAT in male mice, which returns to control levels during activation of orexin neurons (Bunney et al., 2017). A separate study showed that chemogenetic activation of orexin neurons in male and female mice increases SPA and NEAT (Zink et al., 2018) and that repeated stimulation (once per day for 10 days) significantly reduced gains in adiposity in mice fed a high-fat diet. In addition to the direct projection of orexin neurons to several brain sites (Espana, Reis, Valentino, & Berridge, 2005; Estabrooke et al., 2001; Oldfield, Allen, Davern, Giles, & Owens, 2007), recent evidence suggests that orexin neurons activate GABAergic neurons within the LH, which is necessary for changes in physical activity and locomotion caused by stimulation of the orexin neurons (Kosse, Schone, Bracey, & Burdakov, 2017).

Consistent with the findings of the chemogenetic studies, optogenetic activation of the orexin neurons also increases SPA, whereas inhibition of the orexin neurons during a normal active phase reduced SPA (Kotz, Perez-Leighton, Teske, & Billington, 2017). Mice were bred to express ChR2 (stimulatory rhodopsin) or NpHr (inhibitory rhodopsin) in orexin neurons only, and then using either blue light or orange light, optogenetically activated or inhibited orexin neurons, respectively. Optogenetic activation of the orexin neurons significantly increased SPA in mice expressing ChR2, which occurred only during dark cycle periods, whereas inhibition of the orexin neurons reduced SPA (Kotz et al., 2017). These data suggest that the effect of orexin neurons on SPA is dependent on circadian regulation. Additional data showing that inhibition of orexin neurons only decreases SPA during the dark and not the light cycle supports this point (Zink et al., 2018). Furthermore, this concept aligns with prior evidence showing circadian control of the orexin neurons (Estabrooke et al., 2001; Marston et al., 2008; Yoshida et al., 2001), including a reciprocal regulation with SCN (Belle et al., 2014). Together, these data indicate that a physiological function of orexin neurons is to regulate SPA, which is dependent upon time of day and likely reflects circadian influences on the orexin neurons.

Polygenic Models of Variation in SPA: The DIO/DIOR (OP/OR) and HA/LA Rats

A polygenic obesity model, the dietary-induced obese (DIO) and DIO-resistant (DIO-R), has provided several insights into the role of the orexin peptides and their receptors in obesity. These rats, also known as obesity-prone (OP) and obesity-resistant (OR) rats, were derived by outbreeding several generations of high-fat diet-fed Sprague-Dawley (SD) rats selected for higher or similar body weight gain compared to rats fed control diets (Levin, Dunn-Meynell, Balkan, & Keesey, 1997; Levin, Hogan, & Sullivan, 1989; Ricci & Levin, 2003). The OP rats not only showed higher body weight after high-fat diet feeding but also on a normal fat diet (Levin et al., 1989, 1997; Novak et al., 2006; Ricci & Levin, 2003; Teske et al., 2006).

Initial studies were unclear about the influence of SPA on obesity susceptibility in these rats (Levin, 1991), but later studies were conclusive in that the obesity-resistant phenotype of the OR rats is associated with higher levels of SPA (Teske et al., 2006, 2012). Across their lifespan OR rats show higher SPA after orexin-A injection in rostral LH, PVN, or LC compared to OP rats (Novak et al., 2006; Teske et al., 2006; Teske, Perez-Leighton, Billington, & Kotz, 2013), despite similar effects in arousal measured from EEG/EMG implants after injecting orexin-A in rostral LH (Teske et al., 2014). This effect might be mediated by increased orexin receptor expression in the rostral LH in OR compared to OP rats, as these rats had similar levels of prepro-orexin expression in rostral LH (Kotz et al., 2012; Teske et al., 2006). Additional data showed that compared to OP rats, OR rats have higher expression of both orexin receptors mRNA in several brain sites associated with the control of sleep behavior (Teske et al., 2013) as well as other brain sites that regulate SPA, which could also contribute to maintain their overall energy expenditure and protect against weight gain (Mavanji, Teske, Billington, & Kotz, 2010). Together, these data indicate that higher orexin sensitivity due to increased orexin receptor expression in key brain sites of the SPA network may explain higher NEAT and obesity resistance in OR rats compared to OP rats. In a different model of selectively bred rats, Sprague-Dawley rats were bred for either high or low aerobic running capacity based on distance run to fatigue on a treadmill test, resulting in the high-capacity runners (HCR) or low-capacity runners (LCR) (Noland et al., 2007). HCR rats showed higher SPA and energy expenditure compared to LCR (Novak et al., 2009), and like the OR/OP rats, they show higher SPA after orexin-A injection into the PVN (Novak et al., 2010). These data indicate that higher orexin function is associated with SPA and NEAT.

Another model used to explore the role of orexin-mediated SPA and NEAT is the high activity (HA) and low activity (LA) rats. The HA/LA model is generated by measuring 24-h baseline SPA of Sprague-Dawley rats. Similar to OR and HCR rats, HA rats have higher energy expenditure and resistance to obesity caused by high-fat diet consumption as compared to LA rats (Perez-Leighton, Butterick-Peterson, et al., 2013). This model illustrates the influence that SPA can have on body weight due to interindividual variability in SPA, which is also observed in response to caloric restriction or physical exercise (Moretto et al., 2017; Perez-Leighton, Grace, et al., 2014). Analysis of orexin function in the HA/LA rats showed that HA rats have higher expression of orexin mRNA and higher sensitivity to orexin-A injections in rostral LH effects on SPA as compared to that in LA rats (Perez-Leighton, Butterick-Peterson, et al., 2013). This higher sensitivity to orexin-A effects on SPA was not observed after injection in nucleus accumbens shell or in the substantia nigra (Perez-Leighton et al., 2017; Perez-Leighton, Butterick-Peterson, et al., 2013), suggesting again that there are site-specific effects of orexin-A action on SPA. Unlike the OP/OR rats, there were no differences in orexin receptor expression in the rostral LH between HA/LA rats (Perez-Leighton et al., 2017), suggesting that a mechanism other than level of orexin receptors promote higher sensitivity to orexin-A in this model. Together, these results support the concept that higher orexinergic signaling increases resistance to the development of obesity.

SEX DIFFERENCES IN OREXIN FUNCTION IN REGULATION OF SPA

Despite evidence showing that baseline differences in physical activity exist between males and females (Broida & Svare, 1984; Lightfoot, Turner, Daves, Vordermark, & Kleeberger, 2004; Nishi, Ishii, Takahashi, Shiroishi, & Koide, 2010) and that energy balance and weight gain are regulated by estrous cycle (Coborn et al., 2017), the role of orexin peptides on SPA and NEAT in females is largely unknown. Unlike male rats where injection of orexin-A stimulates both ambulation and rearing or vertical movement (Teske et al., 2006), our data show that in gonadally intact female rats, orexin-A in the rostral LH only stimulates ambulation (Fig. 1). Although the contribution of estrous cycle phase to these effects is unclear, a direct comparison between male and female rodents would be required to substantiate the apparent sex effect of orexin-A on rearing.

Sex differences were also noted between gonadally intact male and female mice after mild activation of the hypothalamic orexin-producing neurons with a low, but not high dose, of clozapine n-oxide (Zink et al., 2018). There are sex differences in brain orexin peptide and receptor expression (Brownell & Conti, 2010; Johren, Neidert, Kummer, & Dominiak, 2002; Silveyra, Cataldi, Lux-Lantos, & Libertun, 2009; Taheri, Mahmoodi, Opacka-Juffry, Ghatei, & Bloom, 1999) and the estrous cycle alters brain expression of orexin and its receptors in a brain-site-specific manner (Porkka-Heiskanen, Kalinchuk, Alanko, Huhtaniemi, & Stenberg, 2004; Russell et al., 2001; Wang, Murata, Narita, Honda, & Higuchi, 2003).

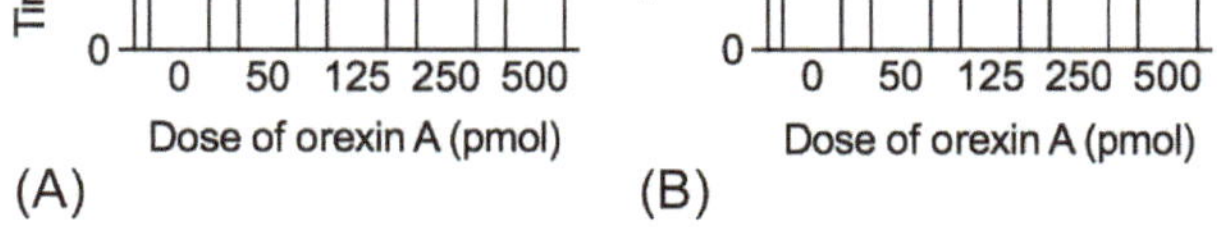

Fig. 1 Orexin-A microinjection into the rostral lateral hypothalamic area significantly increases (A) time ambulatory but not (B) time vertical in gonadally intact female Sprague-Dawley rats (repeated measures ANOVA $F_{(4,\ 32)}=6.7$, $P=.0004$). Data expressed as mean ± SEM. $N=9$. $*P<.05$ as compared to the zero and 50pmol dose of orexin-A. Comparisons between doses corrected for the false discovery rate.

Orexin peptide expression is higher during proestrous (Porkka-Heiskanen et al., 2004), and estrogens modulate the effect of orexin on arousal (Easton, Dwyer, & Pfaff, 2006). Female mice lacking orexin neurons or orexin peptides gain more weight compared to males lacking orexin neurons and female wild-type mice (Fujiki et al., 2006; Ramanathan & Siegel, 2014; Tsuneki et al., 2008). Thus additional work is needed to determine the behavioral significance of the reported neuroanatomical differences in the orexin system between male and female rodents.

OREXIN MECHANISMS FOR INCREASING COMPONENTS OF TOTAL ENERGY EXPENDITURE

Influence on EE During Sleep/Wake States

In addition to their role in promoting SPA and NEAT, there is evidence that orexin peptides can increase other components of energy expenditure. For example, NEAT only accounts for $\approx$17% of the increase in total energy expenditure due to repeated orexin-A injection in rostral LH (Perez-Leighton, Boland, Teske, Billington, & Kotz, 2012). Using more accurate methods of energy expenditure analysis, it was demonstrated that injection of orexin-A in the VLPO increases EE due to increased SPA (Mavanji et al., 2015), but also during rest and sleep; all of which were blocked by pretreatment with the dual orexin receptor antagonist, TCS1102 (Coborn et al., 2017) and by both acute and chronic sleep deprivation (DePorter, Coborn, & Teske, 2017).

Brown Adipose Tissue Thermogenesis

Several lines of evidence suggest that orexin can increase thermogenesis in brown adipose tissue (BAT) (Perez-Leighton, Billington, and Kotz, 2014). It has been proposed that orexin acts as modulator rather than a driver of BAT thermogenesis, which is controlled in response to changes in temperature by a neuronal circuit whose primary input is changes in environmental temperature (Tupone, Madden, & Morrison, 2014). Lack of orexin peptide expression blunts development of BAT in mice, which is restored by orexin-A injections (Sellayah et al., 2011) and rats with depletion of orexin neurons show blunted BAT increases in response to stressors (Mohammed, Ootsuka, Yanagisawa, & Blessing, 2014). In mice deficient for orexin, it was reported that hypothalamic control of BAT thermogenesis was altered, suggesting that orexin neurons are downstream of other parts of the network that regulates BAT thermogenesis (Martins et al., 2016). Studies have demonstrated that in aged mice (Sellayah & Sikder, 2014) BAT dysfunction and fat mass are augmented in mice lacking orexin neurons whereas orexin injections in old mice restored BAT function and reduced adiposity.

Orexin neurons have collateral efferents to both CNS regions and BAT (Oldfield et al., 2007) and retrograde tracer injections show the sympathetic circuit that controls BAT involves several brain areas innervated by orexin neurons, including the dorsal raphe nucleus and the PVN (Bartness & Song, 2007; Stanley et al., 2010). Intracerebroventricular injection of orexin-A in rats increases sympathetic activity leading to increase interscapular BAT thermogenesis and colonic temperature, which is decreased by pretreatment with haloperidol (Monda, Viggiano, & De Luca, 2003), VMH lesions (Monda et al., 2005), electrochemical lesions of DA neurons with 6-OHDA (Monda et al., 2007), and olanzapine (Monda et al., 2008).

Direct evidence for involvement of orexin in regulation of BAT comes from studies showing that disinhibition of the lateral hypothalamic area increased BAT temperature and sympathetic nervous system activity to BAT (Cerri & Morrison, 2005) and injection of orexin-A in the dorsal raphe and parapyramidal nucleus stimulates BAT thermogenesis (Tupone, Madden, Cano, & Morrison, 2011). However, the translational validity of orexin-A-stimulated increase in BAT thermogenesis shown in rodents has been questioned since a recent report found that orexin-A failed to stimulate thermogenesis in human BAT in vitro (Pino, Divoux, Simmonds, Smith, & Sparks, 2017). These data need to be considered with caution since cells grown outside of their normal regulatory innervations cannot be governed in the same way as cells within their normal physiologically relevant milieu.

CONCLUSION

In conclusion, it is clear that orexin regulates NEAT and is important for body weight regulation. This is based on findings from human and animal work, and many mechanistic studies suggesting that neural systems regulate SPA and NEAT. There may be shared neural systems underlying NEAT and the restless behavior associated with other disorders such as anorexia nervosa, but for obesity prevention this behavior has distinct caloric advantages. Future challenges in this field are finding new ways to noninvasively target NEAT-activating brain sites in humans for the purpose of better body weight regulation.

REFERENCES

Akiyama, M., Yuasa, T., Hayasaka, N., Horikawa, K., Sakurai, T., & Shibata, S. (2004). Reduced food anticipatory activity in genetically orexin (hypocretin) neuron-ablated mice. *The European Journal of Neuroscience*, *20*(11), 3054–3062.

Bartness, T. J., & Song, C. K. (2007). Thematic review series: Adipocyte biology. Sympathetic and sensory innervation of white adipose tissue. *Journal of Lipid Research*, *48*(8), 1655–1672.

Belle, M. D., Hughes, A. T., Bechtold, D. A., Cunningham, P., Pierucci, M., Burdakov, D., et al. (2014). Acute suppressive and long-term phase modulation actions of orexin on the mammalian circadian clock. *The Journal of Neuroscience*, *34*(10), 3607–3621.

Broida, J., & Svare, B. (1984). Sex differences in the activity of mice: Modulation by postnatal gonadal hormones. *Hormones and Behavior*, *18*(1), 65–78.

Brownell, S. E., & Conti, B. (2010). Age- and gender-specific changes of hypocretin immunopositive neurons in C57Bl/6 mice. *Neuroscience Letters*, *472*(1), 29–32.

Bunney, P. E., Zink, A. N., Holm, A. A., Billington, C. J., & Kotz, C. M. (2017). Orexin activation counteracts decreases in nonexercise activity thermogenesis (NEAT) caused by high-fat diet. *Physiology & Behavior*, *176*, 139–148.

Cerri, M., & Morrison, S. F. (2005). Activation of lateral hypothalamic neurons stimulates brown adipose tissue thermogenesis. *Neuroscience*, *135*(2), 627–638.

Coborn, J. E., DePorter, D. P., Mavanji, V., Sinton, C. M., Kotz, C. M., Billington, C. J., et al. (2017). Role of orexin-A in the ventrolateral preoptic area on components of total energy expenditure. *International Journal of Obesity*, *41*(8), 1256–1262.

DePorter, D. P., Coborn, J. E., & Teske, J. A. (2017). Partial sleep deprivation reduces the efficacy of orexin-A to stimulate physical activity and energy expenditure. *Obesity (Silver Spring)*, *25*(10), 1716–1722.

Dube, M. G., Kalra, S. P., & Kalra, P. S. (1999). Food intake elicited by central administration of orexins/hypocretins: Identification of hypothalamic sites of action. *Brain Research*, *842* (2), 473–477.

Easton, A., Dwyer, E., & Pfaff, D. W. (2006). Estradiol and orexin-2 saporin actions on multiple forms of behavioral arousal in female mice. *Behavioral Neuroscience*, *120*(1), 1–9.

Espana, R. A., Reis, K. M., Valentino, R. J., & Berridge, C. W. (2005). Organization of hypocretin/orexin efferents to locus coeruleus and basal forebrain arousal-related structures. *The Journal of Comparative Neurology*, *481*(2), 160–178.

Estabrooke, I. V., McCarthy, M. T., Ko, E., Chou, T. C., Chemelli, R. M., Yanagisawa, M., et al. (2001). Fos expression in orexin neurons varies with behavioral state. *The Journal of Neuroscience*, *21*(5), 1656–1662.

Fruhbeck, G. (2005). Does a NEAT difference in energy expenditure lead to obesity? *Lancet*, *366*(9486), 615–616.

Fujiki, N., Yoshida, Y., Zhang, S., Sakurai, T., Yanagisawa, M., & Nishino, S. (2006). Sex difference in body weight gain and leptin signaling in hypocretin/orexin deficient mouse models. *Peptides*, *27*(9), 2326–2331.

Funato, H., Tsai, A. L., Willie, J. T., Kisanuki, Y., Williams, S. C., Sakurai, T., et al. (2009). Enhanced orexin receptor-2 signaling prevents diet-induced obesity and improves leptin sensitivity. *Cell Metabolism*, *9*(1), 64–76.

Garland, T., Jr., Schutz, H., Chappell, M. A., Keeney, B. K., Meek, T. H., Copes, L. E., et al. (2011). The biological control of voluntary exercise, spontaneous physical activity and daily energy expenditure in relation to obesity: Human and rodent perspectives. *The Journal of Experimental Biology*, *214*(Pt 2), 206–229.

Hagan, J. J., Leslie, R. A., Patel, S., Evans, M. L., Wattam, T. A., Holmes, S., et al. (1999). Orexin A activates locus coeruleus cell firing and increases arousal in the rat. *Proceedings of the National Academy of Sciences of the United States of America*, *96*(19), 10911–10916.

Hara, J., Beuckmann, C. T., Nambu, T., Willie, J. T., Chemelli, R. M., Sinton, C. M., et al. (2001). Genetic ablation of orexin neurons in mice results in narcolepsy, hypophagia, and obesity. *Neuron, 30*(2), 345–354.

Hara, J., Yanagisawa, M., & Sakurai, T. (2005). Difference in obesity phenotype between orexin-knockout mice and orexin neuron-deficient mice with same genetic background and environmental conditions. *Neuroscience Letters, 380*(3), 239–242.

Inutsuka, A., Inui, A., Tabuchi, S., Tsunematsu, T., Lazarus, M., & Yamanaka, A. (2014). Concurrent and robust regulation of feeding behaviors and metabolism by orexin neurons. *Neuropharmacology, 85*, 451–460.

Irukayama-Tomobe, Y., Ogawa, Y., Tominaga, H., Ishikawa, Y., Hosokawa, N., Ambai, S., et al. (2017). Nonpeptide orexin type-2 receptor agonist ameliorates narcolepsy-cataplexy symptoms in mouse models. *Proceedings of the National Academy of Sciences of the United States of America, 114*(22), 5731–5736.

Johren, O., Neidert, S. J., Kummer, M., & Dominiak, P. (2002). Sexually dimorphic expression of prepro-orexin mRNA in the rat hypothalamus. *Peptides, 23*(6), 1177–1180.

Jones, D. N., Gartlon, J., Parker, F., Taylor, S. G., Routledge, C., Hemmati, P., et al. (2001). Effects of centrally administered orexin-B and orexin-A: A role for orexin-1 receptors in orexin-B-induced hyperactivity. *Psychopharmacology, 153*(2), 210–218.

Kiwaki, K., Kotz, C. M., Wang, C., Lanningham-Foster, L., & Levine, J. A. (2004). Orexin A (hypocretin 1) injected into hypothalamic paraventricular nucleus and spontaneous physical activity in rats. *American Journal of Physiology. Endocrinology and Metabolism, 286*(4), E551–E559.

Kosse, C., Schone, C., Bracey, E., & Burdakov, D. (2017). Orexin-driven GAD65 network of the lateral hypothalamus sets physical activity in mice. *Proceedings of the National Academy of Sciences of the United States of America, 114*(17), 4525–4530.

Kotz, C., Nixon, J., Butterick, T., Perez-Leighton, C., Teske, J., & Billington, C. (2012). Brain orexin promotes obesity resistance. *Annals of the New York Academy of Sciences, 1264*, 72–86.

Kotz, C. M., & Levine, J. A. (2005). Role of nonexercise activity thermogenesis (NEAT) in obesity. *Minnesota Medicine, 88*(9), 54–57.

Kotz, C. M., Perez-Leighton, C. E., Teske, J. A., & Billington, C. J. (2017). Spontaneous physical activity defends against obesity. *Current Obesity Reports, 6*(4), 362–370.

Kotz, C. M., Teske, J. A., & Billington, C. J. (2008). Neuroregulation of nonexercise activity thermogenesis and obesity resistance. *American Journal of Physiology. Regulatory, Integrative and Comparative Physiology, 294*(3), R699–R710.

Kotz, C. M., Teske, J. A., Levine, J. A., & Wang, C. (2002). Feeding and activity induced by orexin A in the lateral hypothalamus in rats. *Regulatory Peptides, 104*(1-3), 27–32.

Kotz, C. M., Wang, C., Teske, J. A., Thorpe, A. J., Novak, C. M., Kiwaki, K., et al. (2006). Orexin A mediation of time spent moving in rats: Neural mechanisms. *Neuroscience, 142*(1), 29–36.

Levin, B. E. (1991). Spontaneous motor activity during the development and maintenance of diet-induced obesity in the rat. *Physiology & Behavior, 50*(3), 573–581.

Levin, B. E., Dunn-Meynell, A. A., Balkan, B., & Keesey, R. E. (1997). Selective breeding for diet-induced obesity and resistance in Sprague-Dawley rats. *The American Journal of Physiology, 273*(2 Pt 2), R725–R730.

Levin, B. E., Hogan, S., & Sullivan, A. C. (1989). Initiation and perpetuation of obesity and obesity resistance in rats. *The American Journal of Physiology, 256*(3 Pt 2), R766–R771.

Levine, J. A. (2002). Non-exercise activity thermogenesis (NEAT). *Best Practice & Research. Clinical Endocrinology & Metabolism, 16*(4), 679–702.

Levine, J. A., Eberhardt, N. L., & Jensen, M. D. (1999). Role of nonexercise activity thermogenesis in resistance to fat gain in humans. *Science, 283*(5399), 212–214.

Levine, J. A., Lanningham-Foster, L. M., McCrady, S. K., Krizan, A. C., Olson, L. R., Kane, P. H., et al. (2005). Interindividual variation in posture allocation: Possible role in human obesity. *Science, 307*(5709), 584–586.

Levine, J. A., Vander Weg, M. W., Hill, J. O., & Klesges, R. C. (2006). Non-exercise activity thermogenesis: The crouching tiger hidden dragon of societal weight gain. *Arteriosclerosis, Thrombosis, and Vascular Biology, 26*(4), 729–736.

Lightfoot, J. T., Turner, M. J., Daves, M., Vordermark, A., & Kleeberger, S. R. (2004). Genetic influence on daily wheel running activity level. *Physiological Genomics, 19*(3), 270–276.

Marston, O. J., Williams, R. H., Canal, M. M., Samuels, R. E., Upton, N., & Piggins, H. D. (2008). Circadian and dark-pulse activation of orexin/hypocretin neurons. *Molecular Brain, 1*, 19.

Martins, L., Seoane-Collazo, P., Contreras, C., Gonzalez-Garcia, I., Martinez-Sanchez, N., Gonzalez, F., et al. (2016). A functional link between AMPK and orexin mediates the effect of BMP8B on energy balance. *Cell Reports, 16*(8), 2231–2242.

Mavanji, V., Perez-Leighton, C. E., Kotz, C. M., Billington, C. J., Parthasarathy, S., Sinton, C. M., et al. (2015). Promotion of wakefulness and energy expenditure by orexin-A in the ventrolateral preoptic area. *Sleep, 38*(9), 1361–1370.

Mavanji, V., Teske, J. A., Billington, C. J., & Kotz, C. M. (2010). Elevated sleep quality and orexin receptor mRNA in obesity-resistant rats. *International Journal of Obesity, 34*(11), 1576–1588.

Mieda, M., Willie, J. T., Hara, J., Sinton, C. M., Sakurai, T., & Yanagisawa, M. (2004). Orexin peptides prevent cataplexy and improve wakefulness in an orexin neuron-ablated model of narcolepsy in mice. *Proceedings of the National Academy of Sciences of the United States of America, 101*(13), 4649–4654.

Mohammed, M., Ootsuka, Y., Yanagisawa, M., & Blessing, W. (2014). Reduced brown adipose tissue thermogenesis during environmental interactions in transgenic rats with ataxin-3-mediated ablation of hypothalamic orexin neurons. *American Journal of Physiology: Regulatory, Integrative and Comparative Physiology, 307*(8), R978–R989.

Monda, M., Viggiano, A., & De Luca, V. (2003). Haloperidol reduces the sympathetic and thermogenic activation induced by orexin A. *Neuroscience Research, 45*(1), 17–23.

Monda, M., Viggiano, A., Viggiano, A., Mondola, R., Viggiano, E., Messina, G., et al. (2008). Olanzapine blocks the sympathetic and hyperthermic reactions due to cerebral injection of orexin A. *Peptides, 29*(1), 120–126.

Monda, M., Viggiano, A., Viggiano, A., Viggiano, E., Messina, G., Tafuri, D., et al. (2007). Sympathetic and hyperthermic reactions by orexin A: Role of cerebral catecholaminergic neurons. *Regulatory Peptides, 139*(1-3), 39–44.

Monda, M., Viggiano, A. N., Viggiano, A., Viggiano, E., Lanza, A., & De Luca, V. (2005). Hyperthermic reactions induced by orexin A: Role of the ventromedial hypothalamus. *The European Journal of Neuroscience, 22*(5), 1169–1175.

Moretto, T. L., Benfato, I. D., de Carvalho, F. P., Barthichoto, M., Le Sueur-Maluf, L. , & de Oliveira, C. A. M. (2017). The effects of calorie-matched high-fat diet consumption on spontaneous physical activity and development of obesity. *Life Sciences, 179*, 30–36.

Nishi, A., Ishii, A., Takahashi, A., Shiroishi, T., & Koide, T. (2010). QTL analysis of measures of mouse home-cage activity using B6/MSM consomic strains. *Mammalian Genome, 21*(9-10), 477–485.

Noland, R. C., Thyfault, J. P., Henes, S. T., Whitfield, B. R., Woodlief, T. L., Evans, J. R., et al. (2007). Artificial selection for high-capacity endurance running is protective against high-fat diet-induced insulin resistance. *American Journal of Physiology. Endocrinology and Metabolism, 293*(1), E31–E41.

Novak, C. M., Escande, C., Burghardt, P. R., Zhang, M., Barbosa, M. T., Chini, E. N., et al. (2010). Spontaneous activity, economy of activity, and resistance to diet-induced obesity in rats bred for high intrinsic aerobic capacity. *Hormones and Behavior, 58*(3), 355–367.

Novak, C. M., Escande, C., Gerber, S. M., Chini, E. N., Zhang, M., Britton, S. L., et al. (2009). Endurance capacity, not body size, determines physical activity levels: Role of skeletal muscle PEPCK. *PLoS One, 4*(6), e5869.

Novak, C. M., Kotz, C. M., & Levine, J. A. (2006). Central orexin sensitivity, physical activity, and obesity in diet-induced obese and diet-resistant rats. *American Journal of Physiology. Endocrinology and Metabolism, 290*(2), E396–E403.

Novak, C. M., & Levine, J. A. (2009). Daily intraparaventricular orexin-A treatment induces weight loss in rats. *Obesity (Silver Spring), 17*(8), 1493–1498.

Oldfield, B. J., Allen, A. M., Davern, P., Giles, M. E., & Owens, N. C. (2007). Lateral hypothalamic 'command neurons' with axonal projections to regions involved in both feeding and thermogenesis. *The European Journal of Neuroscience, 25*(8), 2404–2412.

Perez-Leighton, C., Little, M. R., Grace, M., Billington, C., & Kotz, C. M. (2017). Orexin signaling in rostral lateral hypothalamus and nucleus accumbens shell in the control of spontaneous physical activity in high- and low-activity rats. *American Journal of Physiology. Regulatory, Integrative and Comparative Physiology, 312*(3), R338–r346.

Perez-Leighton, C. E., Billington, C. J., & Kotz, C. M. (2014). Orexin modulation of adipose tissue. *Biochimica et Biophysica Acta, 1842*(3), 440–445.

Perez-Leighton, C. E., Boland, K., Billington, C. J., & Kotz, C. M. (2013). High and low activity rats: elevated intrinsic physical activity drives resistance to diet-induced obesity in non-bred rats. *Obesity (Silver Spring), 21*(2), 353–360.

Perez-Leighton, C. E., Boland, K., Teske, J. A., Billington, C., & Kotz, C. M. (2012). Behavioral responses to orexin, orexin receptor gene expression, and spontaneous physical activity contribute to individual sensitivity to obesity. *American Journal of Physiology. Endocrinology and Metabolism, 303*(7), E865–E874.

Perez-Leighton, C. E., Butterick-Peterson, T. A., Billington, C. J., & Kotz, C. M. (2013). Role of orexin receptors in obesity: From cellular to behavioral evidence. *International Journal of Obesity, 37*(2), 167–174.

Perez-Leighton, C. E., Grace, M., Billington, C. J., & Kotz, C. M. (2014). Role of spontaneous physical activity in prediction of susceptibility to activity based anorexia in male and female rats. *Physiology & Behavior, 135*, 104–111.

Pino, M. F., Divoux, A., Simmonds, A. V., Smith, S. R., & Sparks, L. M. (2017). Investigating the effects of orexin-A on thermogenesis in human deep neck brown adipose tissue. *International Journal of Obesity, 41*(11), 1646–1653.

Porkka-Heiskanen, T., Kalinchuk, A., Alanko, L., Huhtaniemi, I., & Stenberg, D. (2004). Orexin A and B levels in the hypothalamus of female rats: The effects of the estrous cycle and age. *European Journal of Endocrinology, 150*(5), 737–742.

Ramanathan, L., & Siegel, J. M. (2014). Gender differences between hypocretin/orexin knockout and wild type mice: age, body weight, body composition, metabolic markers, leptin and insulin resistance. *Journal of Neurochemistry, 131*(5), 615–624.

Ravussin, E. (2005). Physiology. A NEAT way to control weight? *Science, 307*(5709), 530–531.

Ravussin, E., Lillioja, S., Anderson, T. E., Christin, L., & Bogardus, C. (1986). Determinants of 24-hour energy expenditure in man. Methods and results using a respiratory chamber. *The Journal of Clinical Investigation, 78*(6), 1568–1578.

Ricci, M. R., & Levin, B. E. (2003). Ontogeny of diet-induced obesity in selectively bred Sprague-Dawley rats. *American Journal of Physiology. Regulatory, Integrative and Comparative Physiology, 285*(3), R610–R618.

Russell, S. H., Small, C. J., Kennedy, A. R., Stanley, S. A., Seth, A., Murphy, K. G., et al. (2001). Orexin A interactions in the hypothalamo-pituitary gonadal axis. *Endocrinology, 142*(12), 5294–5302.

Sellayah, D., Bharaj, P., & Sikder, D. (2011). Orexin is required for brown adipose tissue development, differentiation, and function. *Cell Metabolism, 14*(4), 478–490.

Sellayah, D., & Sikder, D. (2014). Orexin restores aging-related brown adipose tissue dysfunction in male mice. *Endocrinology, 155*(2), 485–501.

Silveyra, P., Cataldi, N. I., Lux-Lantos, V., & Libertun, C. (2009). Gonadal steroids modulated hypocretin/orexin type-1 receptor expression in a brain region, sex and daytime specific manner. *Regulatory Peptides, 158*(1-3), 121–126.

Snitker, S., Tataranni, P. A., & Ravussin, E. (2001). Spontaneous physical activity in a respiratory chamber is correlated to habitual physical activity. *International Journal of Obesity and Related Metabolic Disorders, 25*(10), 1481–1486.

Stanley, S., Pinto, S., Segal, J., Perez, C. A., Viale, A., DeFalco, J., et al. (2010). Identification of neuronal subpopulations that project from hypothalamus to both liver and adipose tissue polysynaptically. *Proceedings of the National Academy of Sciences of the United States of America, 107*(15), 7024–7029.

Sweet, D. C., Levine, A. S., Billington, C. J., & Kotz, C. M. (1999). Feeding response to central orexins. *Brain Research, 821*(2), 535–538.

Taheri, S., Mahmoodi, M., Opacka-Juffry, J., Ghatei, M. A., & Bloom, S. R. (1999). Distribution and quantification of immunoreactive orexin A in rat tissues. *FEBS Letters, 457* (1), 157–161.

Teske, J. A., Billington, C. J., & Kotz, C. M. (2008). Neuropeptidergic mediators of spontaneous physical activity and non-exercise activity thermogenesis. *Neuroendocrinology, 87* (2), 71–90.

Teske, J. A., Billington, C. J., & Kotz, C. M. (2014). Mechanisms underlying obesity resistance associated with high spontaneous physical activity. *Neuroscience, 256*, 91–100.

Teske, J. A., Billington, C. J., Kuskowski, M. A., & Kotz, C. M. (2012). Spontaneous physical activity protects against fat mass gain. *International Journal of Obesity, 36*(4), 603–613.

Teske, J. A., Levine, A. S., Kuskowski, M., Levine, J. A., & Kotz, C. M. (2006). Elevated hypothalamic orexin signaling, sensitivity to orexin A, and spontaneous physical activity in obesity-resistant rats. *American Journal of Physiology. Regulatory, Integrative and Comparative Physiology, 291*(4), R889–R899.

Teske, J. A., Perez-Leighton, C. E., Billington, C. J., & Kotz, C. M. (2013). Role of the locus coeruleus in enhanced orexin A-induced spontaneous physical activity in obesity-resistant rats. *American Journal of Physiology. Regulatory, Integrative and Comparative Physiology, 305*(11), R1337–R1345.

Thorpe, A. J., & Kotz, C. M. (2005). Orexin A in the nucleus accumbens stimulates feeding and locomotor activity. *Brain Research, 1050*(1-2), 156–162.

Tsuneki, H., Murata, S., Anzawa, Y., Soeda, Y., Tokai, E., Wada, T., et al. (2008). Age-related insulin resistance in hypothalamus and peripheral tissues of orexin knockout mice. *Diabetologia, 51*(4), 657–667.

Tupone, D., Madden, C. J., Cano, G., & Morrison, S. F. (2011). An orexinergic projection from perifornical hypothalamus to raphe pallidus increases rat brown adipose tissue thermogenesis. *The Journal of Neuroscience, 31*(44), 15944–15955.

Tupone, D., Madden, C. J., & Morrison, S. F. (2014). Autonomic regulation of brown adipose tissue thermogenesis in health and disease: Potential clinical applications for altering BAT thermogenesis. *Frontiers in Neuroscience, 8*, 14.

Wang, J. B., Murata, T., Narita, K., Honda, K., & Higuchi, T. (2003). Variation in the expression of orexin and orexin receptors in the rat hypothalamus during the estrous cycle, pregnancy, parturition, and lactation. *Endocrine, 22*(2), 127–134.

Winrow, C. J., Gotter, A. L., Cox, C. D., Doran, S. M., Tannenbaum, P. L., Breslin, M. J., et al. (2011). Promotion of sleep by suvorexant-a novel dual orexin receptor antagonist. *Journal of Neurogenetics*, *25*(1-2), 52–61.

Winrow, C. J., Gotter, A. L., Cox, C. D., Tannenbaum, P. L., Garson, S. L., Doran, S. M., et al. (2012). Pharmacological characterization of MK-6096—A dual orexin receptor antagonist for insomnia. *Neuropharmacology*, *62*(2), 978–987.

Yoshida, Y., Fujiki, N., Nakajima, T., Ripley, B., Matsumura, H., Yoneda, H., et al. (2001). Fluctuation of extracellular hypocretin-1 (orexin A) levels in the rat in relation to the light-dark cycle and sleep-wake activities. *The European Journal of Neuroscience*, *14*(7), 1075–1081.

Zink, A. N., Bunney, P. E., Holm, A. A., Billington, C. J., & Kotz, C. M. (2018). Neuromodulation of orexin neurons reduces diet-induced adiposity. *International Journal of Obesity*, *42*(4), 737–745.

Zurlo, F., Ferraro, R. T., Fontvielle, A. M., Rising, R., Bogardus, C., & Ravussin, E. (1992). Spontaneous physical activity and obesity: Cross-sectional and longitudinal studies in Pima Indians. *The American Journal of Physiology*, *263*(2 Pt 1), E296–E300.

CHAPTER 4

Orexins and Control of Feeding by Learned Cues

Gorica D. Petrovich
Department of Psychology, Boston College, Chestnut Hill, MA, United States

INTRODUCTION

Feeding is a motivated behavior necessary for survival. It is comprised of food seeking and consumption and these behaviors are coordinated with accompanying cognitive and bodily (metabolic, autonomic, and visceral) responses. The control of feeding behavior therefore necessitates an integrated network across behavioral, cognitive, and physiological neural systems (Swanson, 2000). That complex neural network mediates the expression of feeding behavior via innate and learned mechanisms, in response to internal and external signals—physiological signals related to energy and nutrient needs and environmental signals that function through hedonic and cognitive processes to drive appetite and eating.

Innate, hard-wired mechanisms can overtake the control of feeding behavior when a fast (reflex-type) response is required, for instance, when an organism's survival is threatened. Under most other circumstances a more complex process underlies the control of feeding behavior. That process necessitates continuous assessments of available energy and nutrients against anticipated usage and gains, which require learning and memory and other cognitive processes (e.g., planning, decision-making) to guide feeding behavior. Learning and memory provide adaptive control of feeding behavior that is important in health and disease and recent research has begun deciphering how these cognitive processes are integrated within the feeding neural network.

A main node of the feeding network, the lateral hypothalamus (lateral hypothalamic area (LHA); Swanson, 2004) was recently proposed to serve as a cognition-motivation interface in the control of feeding behavior (Petrovich, 2018) and the neuropeptides orexins/hypocretins are among

The Orexin/Hypocretin System
https://doi.org/10.1016/B978-0-12-813751-2.00004-8

key LHA substrates mediating this function. This chapter will review the evidence for a causal function of the orexin system in the control of feeding behavior by learned cues—a model of cognitive motivation to eat.

OREXIN NEURONS: A BRIEF HISTORY AND ANATOMICAL ORGANIZATION

The orexin/hypocretin (ORX) neuropeptides were discovered, named, and characterized independently by two research groups in the late 1990s (de Lecea et al., 1998; Sakurai et al., 1998). Early work demonstrated that they are important for wakefulness and for regulation of feeding behavior (Chemelli et al., 1999; de Lecea et al., 1998; Sakurai et al., 1998). Since then, substantial evidence has been accumulated to support multiple ORX functions in behavioral state control and motivation underlying feeding and other reward behaviors (Boutrel, Cannella, & de Lecea, 2010; Hurley & Johnson, 2014; Mahler, Moorman, Smith, James, & Aston-Jones, 2014; Scammell, Arrigoni, & Lipton, 2017). A unifying function across these different processes has been proposed that ORX mediates the integration between motivation and behavior—translating motivation into action (Mahler et al., 2014). Related to that function, the evidence is reviewed in this chapter that ORX may integrate cognitive processes with behavioral motivation (and arousal) in the context of feeding behavior.

ORX is exclusively produced by anatomically circumscribed group of neurons that are concentrated mainly within the LHA and are distinct from the neurons expressing another orexigenic neuropeptide, melanin-concentrating hormone (de Lecea et al., 1998; Elias et al., 1998; Sakurai et al., 1998; Swanson, Sanchez-Watts, & Watts, 2005). There are two forms of the ORX peptide, Orexin-A and Orexin-B (Hypocretin 1 and 2), that are cleaved from the precursor, prepro-orexin (de Lecea et al., 1998; Sakurai et al., 1998). To mediate their functions, ORX-A and ORX-B signal via two types of G protein-coupled receptors (1 and 2), which they bind with different affinities (Marcus et al., 2001; Sakurai et al., 1998; Scammell & Winrow, 2011).

In contrast to the confined location of the cell bodies of ORX neurons, ORX fibers and receptors are very broadly distributed across the forebrain-brainstem neural axis (Marcus et al., 2001; Peyron et al., 1998) and overlap with dopamine inputs (Baldo, Daniel, Berridge, & Kelley, 2003). That structural organization enables ORXs to exert potentially widespread influence across different functional systems and across different stages of information processing—from sensory inputs to behavioral outputs. The ability to influence both behavior and cognition is in accordance with the importance of ORXs in behavioral state control (wakefulness and arousal), reward processing, and motivation (Boutrel et al., 2010; Hurley & Johnson, 2014; Mahler et al., 2014; Sakurai, 2014).

In addition to the ability to influence diverse systems, ORX neurons can have multifaceted effects at their targets depending whether they are coreleased with other neurotransmitters or peptides. ORX neurons are not homogenous; they produce additional neurotransmitters and neuropeptides, which when released with ORX can act in concomitant or opposing ways (Li & van den Pol, 2006; Muschamp et al., 2014; for reviews see Bonnavion, Mickelsen, Fujita, de Lecea, & Jackson, 2016; Schöne & Burdakov, 2012). Most ORX neurons are glutamatergic, as they coexpress vesicular glutamate transporter VGLUT1 or 2 but some may be GABAergic, as they have been shown to express GAD1 (Mickelsen et al., 2017; Rosin, Weston, Sevigny, Stornetta, & Guyenet, 2003). ORX neurons also express opioid peptide dynorphin and other peptides and neuromodulators (for a recent review, see Bonnavion et al., 2016). Additional diversity of ORX neurons is marked by inputs from LHA neurons with receptors for adipose-produced hormone leptin. ORX neurons do not express the long form of the leptin receptor (LepRb) but receive inputs from a subpopulation of LepRb LHA neurons that coexpress neurotensin (Leinninger et al., 2011, for review, see Brown, Woodworth, & Leinninger, 2015).

The organization of ORX neurons in regard to how they are integrated within the LHA and whether distinct subpopulations could be distinguished based on theirs inputs and outputs are among the most pressing current research questions. It is becoming increasingly clear that multiple LHA substrates are required to mediate coordinated control of feeding behavior and ORX neurons are an important part of that puzzle. The chemo- and optogenetic methods are beginning to uncover the LHA cell-specific organization. These methods have enabled selective interrogations of specific, genetically defined neurons and their pathways, including manipulations during temporally precise events in behaving animals (Fenno, Yizhar, & Deisseroth, 2011; Sternson & Roth, 2014). Notably, glutamatergic LHA neurons that possibly included ORX neurons, and GABAergic LHA neurons have been shown to regulate feeding in opposite directions. Optogenetic stimulation of glutamatergic (VGlut2-expressing) LHA neurons suppressed while their inhibition initiated feeding (Jennings, Rizzi, Stamatakis, Ung, & Stuber, 2013). Similarly, genetic ablation of these neurons increased food intake and weight gain in mice fed high-fat diet, without changes in general locomotion (Stamatakis et al., 2016). Conversely, photoactivation of the LHA GABA neurons stimulated, while their inhibition halted feeding (Jennings et al., 2013; O'Connor et al., 2015). Nevertheless, how optogenetic manipulations relate to naturally occurring physiological events is not fully understood and there is uncertainty regarding which substrates and neuronal subpopulations mediate feeding behavior specifically versus general motor pattern output for ingestion (Navarro et al., 2016).

OREXIN SYSTEM AND FEEDING BEHAVIOR

Initial evidence that the ORX system is important in the control of feeding behavior came from the findings that ORX mRNA is upregulated by fasting and when exogenously applied ORXs stimulated food consumption (Haynes et al., 1999; Sakurai et al., 1998; Sweet, Levine, Billington, & Kotz, 1999). Early studies also found that these stimulatory effects on food consumption were blocked with systemic administration of an ORX receptor 1 (ORX-R1) antagonist (Haynes et al., 2000). Subsequent studies found that ORX-R1 signaling mediates operant responding to obtain palatable foods and other rewards (drugs) and is needed when hedonic (taste) or cognitive mechanisms drive feeding behavior (Borgland et al., 2009; Cason & Aston-Jones, 2013; Choi, Davis, Fitzerald, & Benoit, 2010; Cole, Mayer, & Petrovich, 2015; Harris, Wimmer, & Aston-Jones, 2005; Nair, Golden, & Shaham, 2008; Sharf et al., 2010; Thorpe, Cleary, Levine, & Kotz, 2005; for reviews, see Cason et al., 2010; Sharf, Sarhan, & DiLeone, 2010). Studies with ORX knockout (KO) mice also demonstrated that the ORX system supports the motivation to obtain palatable food, as ORX KO mice consumed less sucrose compared to their wild-type littermates (Matsuo et al., 2011).

In regard to cognitive control of feeding, it is now well established, based on findings across studies with different behavioral paradigms, that presentations of food or drug reward predictive cues strongly activate ORX neurons (Choi et al., 2010; Harris et al., 2005; Hassani, Krause, Mainville, Cordova, & Jones, 2016; Petrovich, Hobin, & Reppucci, 2012). Initial evidence came from studies with conditioned place preference tasks, which found Fos induction in ORX neurons when animals explored environments associated with food or drug rewards (Harris et al., 2005). Subsequent studies found that food-associated contextual (feeding environment) and discrete (e.g., tone) cues induced Fos in ORX neurons, typically within the perifornical area (Choi et al., 2010; Petrovich et al., 2012). The Fos induction patterns in these studies were presumed to indicate increased neuronal activation and recent recordings from ORX neurons with the juxtacellular method validate those interpretations (Hassani et al., 2016). Hassani, Jones and colleagues found that ORX neurons discriminated between cues associated with appetitive versus aversive outcomes and fired at significantly higher rates to a reward cue predicting sweet liquid compared to another cue predicting aversive, bitter liquid (Hassani et al., 2016). The evidence that the recruitment of ORX neurons by reward cues reflects a causal function of ORX system signaling via Receptor 1 in cue-driven control of feeding behaviors and during cue–food learning acquisition is discussed next.

Control of Feeding by Learned Cues: A Model of Cognitive Motivation to Eat

In addition to physiological needs, the motivation to seek and consume food can be driven by environmental signals that function through hedonic and cognitive mechanisms. Behavioral models of cognitive motivation include paradigms that use learning and memory to manipulate food seeking (e.g., conditioned place preference, instrumental/operant responding) and consumption (e.g., cue-induced feeding). Typically there are two phases in these preparations in order to test how food-associated cues control feeding behavior. The first phase is learning acquisition, when cue-food associations are formed. The second phase is behavioral expression, when the effects of acquired learned cues on feeding behavior are tested, and that phase involves memory recall of the cue and subsequent induction of food motivation. Recent work has demonstrated that the ORX-R1 system is important during both the acquisition and expression of this cognitive motivation to eat. These behavioral preparations and the cognitive drive to eat are relevant to our inability to resist overeating in the modern world. Therefore the ORX system and its neural targets represent potential sites for therapeutic interventions for those who suffer from insatiable appetite and overeating.

Learned cues can powerfully control feeding behavior and help our survival when they function in concert with the physiological control (Woods & Ramsay, 2000). However, food-associated cues can stimulate food seeking and consumption independent of hunger in animals and people (Birch, McPhee, Sullivan, & Johnson, 1989; Cornell, Rodin, & Weingarten, 1989; Weingarten, 1983) and when they persist could cause overeating and obesity. This form of overeating was likely beneficial in the past, when energy resources were scarce and on-demand appetite and opportunistic eating would have been adaptive in anticipation of future shortages; however, it is maladaptive in our current environment. We are surrounded by easily accessible and affordable palatable foods and omnipresent food cues enhance our susceptibility to overeating and body weight gain (Berthoud, 2012; Petrovich, 2013). In support to this prediction, a recent meta-analysis confirmed that food-cue exposure, and associated experience of craving, significantly influence eating behavior and weight gain in humans (Boswell & Kober, 2016). Importantly, individual differences in neural responding to food cues predict long-term weight gain in humans, and specifically the activation of the basolateral amygdala and its connectivity with the LHA were correlated with subsequent weight gain over a year (Sun et al., 2015). The ORX system is an integral part of that neural circuitry and its dysregulation maybe be an important cause of vulnerability to enhanced cognitive motivation to eat.

OREXIN SYSTEM AND CONTROL OF FEEDING BY LEARNED CUES

Recent studies established a causal role for the ORX-R1 system in cognitively driven consumption using a cue-induced feeding preparation. In that preparation, an initially neutral cue from the environment (tone) becomes a signal for food through associative learning, Pavlovian conditioning—repeated presentations immediately prior to food delivery (Weingarten, 1983). Based on that acquired ability, the cue can later stimulate food consumption independently of physiological hunger. Systemic administration of the ORX-R1 antagonist SB-334867 selectively reduced cue-driven consumption in sated rats but had no effect on baseline eating, demonstrating a specific ORX function in the motivation to eat under the learned cue, rather than a general effect on consumption (Cole, Mayer, & Petrovich, 2015).

The same study conducted complementary neural analysis with Fos induction detection to determine the key sites of ORX-R1 action. Multiple forebrain regions were examined, including the core components of the cue-induced feeding circuitry (medial prefrontal cortex, basolateral amygdala, and LHA; Petrovich, 2013) and the paraventricular thalamus where ORX-R1 signaling was previously shown to mediate hedonic eating (Choi et al., 2012). The analysis revealed that the abolishment of cue-induced feeding by the ORX-R1 antagonist was accompanied by increased Fos expression selectively in the medial prefrontal cortex and the paraventricular thalamus, suggesting they are key brain sites of this systemic effect. These findings point to the medial prefrontal cortex and the paraventricular thalamus as neural substrates where ORX signaling may promote eating under the learned cues and indicate ORX-R1 antagonism as a potential pharmacological target for treatment of disordered eating in humans.

Importantly, in addition to its necessity to stimulate food seeking and consumption under learned cues, ORX neurons are recruited during cue-food associative learning and ORX-R1 signaling modulates the acquisition and extinction of cue-food associations (Cole, Hobin, & Petrovich, 2015; Keefer, Cole, & Petrovich, 2016). Systemic administration of the ORX-R1 antagonist, SB-334867 attenuated the acquisition and extinction of Pavlovian cue-food conditioning (Keefer et al., 2016). This function is not limited to associative food reward learning, as ORX has been shown to mediate other forms of appetitive and aversive learning, including taste preference learning (Mediavilla, Cabello, & Risco, 2011) and acquisition and recall of conditioned place preference for drug rewards (Harris et al., 2005; reviewed in Keefer et al., 2016). For example, ORX KO mice are impaired in spatial working memory (Dang et al., 2018), and ORX treatment restores hippocampal-dependent memory (two-way active avoidance task) in ORX KO mice (Mavanji et al., 2017). Similarly, conditioned place preference induced by chemical stimulation of the LHA requires ORX-R1 signaling in the hippocampal formation (dentate gyrus) during the acquisition, expression, and extinction of this task (Parsania, Moradi, Fatahi, & Haghparast, 2016). Thus similar to cue–food learning (Keefer et al., 2016), in other tasks ORX-R1 signaling is needed across different stages of learning and memory acquisition and extinction. Collectively, these results suggest that the ORX-R1 system may be recruited to enhance learning and memory, and likely other cognitive processes, to support motivated behaviors.

ORX System in Control of Feeding by Learned Cues: Translating Cognitive Processes into Feeding Motivation?

The activation of ORX neurons by food cues alone, in the absence of hunger or food, indicates that these neurons are engaged during a memory recall of the food (and feeding) associated with the cue. That memory induction is an essential component of the process that stimulates feeding (see earlier discussion) and ORX in that setting may be required to link the memory and arousal to induce the underlying drive to eat. In support of this premise, the patterns of ORX neurons' responses suggest that they discharge as reward neurons and as arousal neurons (Hassani et al., 2016). These results were interpreted to indicate that ORX neurons are engaged in reward processes while modulating cortical and behavioral outputs to maintain arousal during learned adaptive behavioral responses (Hassani et al., 2016). Similarly, ORX may maintain arousal while linking cognition and behavior during feeding under learned cues. This function would be in accordance with ORX engagement when heightened motivational state is needed to support adaptive behaviors (Mahler et al., 2014; Sakurai, 2014) and in agreement with prior evidence for its dual role in arousal and energy metabolism. ORX neurons are needed for arousal associated with changes in energy balance and feeding (González et al., 2016; Yamanaka et al., 2003) and sleep and energy metabolism integration (Brown et al., 2015; Saper, 2006).

The neural sites where ORX may mediate cognitive motivation to eat very likely include multiple coordinated areas. The paraventricular thalamus and medial prefrontal cortex were strongly implicated in ORX effects on cue-induced feeding, based on their selective recruitment when systemic blockade of ORX-R1 signaling impaired cue-driven consumption (Cole, Mayer, & Petrovich, 2015; see earlier discussion). Previously, ORX signaling via its receptor 2 was shown to robustly excite the paraventricular thalamus neurons that project to the medial prefrontal cortex, and that cascade was postulated to enhance cognitive arousal, via a feedforward mechanism of direct ORX excitation of the medial prefrontal cortex neurons (Huang, Ghosh, & van den Pol, 2006). Thus a parallel ORX-R1 mechanism may be engaged to enhance cognitive motivation for food and drug rewards. The paraventricular thalamus and medial prefrontal cortex are reciprocally connected (Berendse & Groenewegen, 1991; Li & Kirouac, 2012) and both areas receive direct inputs from LHA and ORX fibers and receptors are particularly dense in the paraventricular thalamus (Hahn & Swanson, 2010; Marcus et al., 2001; Peyron et al., 1998). Thus the two regions could form a functional loop that would further enhance ORX effects, as previously proposed (Cole, Mayer, & Petrovich, 2015).

In addition to bidirectional interactions with the hypothalamic feeding circuitry (Horvath, Diano, & van den Pol, 1999; Jain, Horvath, Kalra, & Kalra, 2000; Yamanaka et al., 2000), the ORX have been shown to influence multiple reward and cognitive processing areas. Those regions include the ventral tegmental area (Baimel, Lau, Qiao, & Borgland, 2017; Fadel & Deutch, 2002; Harris et al., 2005; Zheng, Patterson, & Berthoud, 2007; for a recent review, see Tyree & de Lecea, 2017), ventral striatum and pallidum (Ho & Berridge, 2013; Thorpe & Kotz, 2005), orbitofrontal and insular cortical areas (Castro & Berridge, 2017), and the hippocampal formation (Parsania et al., 2016), as well as feeding hindbrain regions (e.g., the nucleus of the solitary tract, Kay, Parise, Lilly, & Williams, 2014). In turn, the LHA region where ORX neurons are concentrated receives highly organized inputs from the amygdala, medial prefrontal cortex, and hippocampal formation (Hahn & Swanson, 2010; Reppucci & Petrovich, 2016) and these inputs have been shown to target ORX neurons (Sakurai et al., 2005; Yoshida, McCormack, España, Crocker, & Scammell, 2006). For example, stimulation of the medial prefrontal cortex μ-opioid system drives feeding behavior via the LHA and engages ORX neurons (Mena, Selleck, & Baldo, 2013). The physiological hunger signal, ghrelin drives meal-entrained consumption via the ventral hippocampus, which acts via ORX neurons and downstream ORX-R1 receptors (Hsu et al., 2015). Notably, these neural networks are overlapping with the drug reward processing networks, including sites of ORX control (for recent reviews, see James, Mahler, Moorman, & Aston-Jones, 2017; Moorman, 2018).

CONCLUDING REMARKS

The evidence reviewed here suggests that the ORX-R1 system is an important mediator of cognitive motivation to eat and that its dysregulation could lead to enhanced appetites. Of particular importance for potential treatment is selectivity of the observed effects; ORX-R1 manipulations impacted eating under the learned cues but did not affect baseline consumption (Cole, Mayer, & Petrovich, 2015). This is in agreement with prior evidence that ORX ablation in transgenic mice did not impact food intake and body weight (Mieda et al., 2005). Furthermore, within the context of drug addiction, it was proposed that manipulation of ORX-R1 signaling could be utilized to selectively target cue-induced drug cravings while minimally affecting baseline responding to rewards (James et al., 2017). Thus the ORX-R1 system is a promising target for therapeutic treatment for overeating driven by hedonic and cognitive influences.

REFERENCES

Baimel, C., Lau, B. K., Qiao, M., & Borgland, S. L. (2017). Projection-target defined effects of orexin and dynorphin on VTA dopamine neurons. *Cell Reports*, *18*, 1346–1355.
Baldo, B. A., Daniel, R. A., Berridge, C. W., & Kelley, A. E. (2003). Overlaping distribution of orexin/hypocretin- and dopamine-b-hydroxylase immunoreactive fibers in rat brain

regions mediating arousal, motivation, and stress. *Journal of Comparative Neurology, 464*, 220–237.

Berendse, H. W., & Groenewegen, H. J. (1991). Restricted cortical termination fields of the midline and intralaminar thalamic nuclei in the rat. *Neuroscience, 42*, 73–102.

Berthoud, H.-R. (2012). The neurobiology of food intake in an obesogenic environment. *The Proceedings of the Nutrition Society, 71*, 478–487.

Birch, L. L., McPhee, L., Sullivan, S., & Johnson, S. (1989). Conditioned meal initiation in young children. *Appetite, 13*, 105–113.

Bonnavion, P., Mickelsen, L. E., Fujita, A., de Lecea, L., & Jackson, A. C. (2016). Hubs and spokes of the lateral hypothalamus: Cell types, circuits and behaviour. *The Journal of Physiology, 594*(22), 6443–6462.

Borgland, S. L., Chang, S. J., Bowers, M. S., Thompson, J. L., Vittoz, N., Floresco, S. B., et al. (2009). Orexin A/hypocretin-1 selectively promotes motivation for positive reinforcers. *The Journal of Neuroscience, 29*, 11215–11225.

Boswell, R., & Kober, H. (2016). Food cue reactivity and craving predict eating and weight gain: A meta-analytic review. *Obesity Reviews, 17*, 159–177.

Boutrel, B., Cannella, N., & de Lecea, L. (2010). The role of hypocretin in driving arousal and goal-oriented behaviors. *Brain Research, 1314*, 103–111.

Brown, J. A., Woodworth, H. L., & Leinninger, G. M. (2015). To ingest or rest? Specialized roles of lateral hypothalamic area neurons in coordinating energy balance. *Frontiers in Systems Neuroscience, 9*, 9.

Cason, A. M., & Aston-Jones, G. (2013). Role of orexin/hypocretin in conditioned sucrose-seeking in rats. *Psychopharmacology, 226*, 155–165.

Cason, A. M., Smith, R. J., Tahsili-Fahadan, P., Moorman, D. E., Sartor, G. C., & Aston-Jones, G. (2010). Role of orexin/hypocretin in reward-seeking and addiction: Implications for obesity. *Physiology & Behavior, 100*, 419–428.

Castro, D. C., & Berridge, C. W. (2017). Opioid and orexin hedonic hotspots in rat orbitofrontal cortex and insula. *Proceedings of the National Academy of Sciences of the United States of America, 114*, E9125–E9134.

Chemelli, R. M., Willie, J. T., Sinton, C. M., Elmquist, J. K., Scammell, T., Lee, G., et al. (1999). Narcolepsy in orexin knockout mice: molecular genetics of sleep regulation. *Cell, 98*, 437–451.

Choi, D. L., Davis, J. F., Fitzerald, M. E., & Benoit, S. C. (2010). The role of orexin-A in food motivation, reward-based feeding behavior and food-induced neuronal activation in rats. *Neuroscience, 167*, 11–20.

Choi, D. L., Davis, J. F., Magrisso, I. J., Fitzerald, M. E., Lipton, J. W., & Benoit, S. C. (2012). Orexin signaling in the paraventricular thalamic nucleus modulates mesolimbic dopamine and hedonic feeding in the rat. *Neuroscience, 210*, 243–248.

Cole, S., Hobin, M. P., & Petrovich, G. D. (2015). Appetitive associative learning recruits a distinct network with cortical, striatal, and hypothalamic regions. *Neuroscience, 286*, 187–202.

Cole, S., Mayer, H. S., & Petrovich, G. D. (2015). Orexin/hypocretin-1 receptor antagonism selectively reduces cue-induced feeding in sated rats and recruits medial prefrontal cortex and thalamus. *Scientific Reports, 5*, 16143.

Cornell, C. E., Rodin, J., & Weingarten, H. (1989). Stimulus-induced eating when satiated. *Physiology & Behavior, 45*, 695–704.

Dang, R., Chen, Q., Song, J., He, C., Zhang, J., Xia, J., et al. (2018). Orexin knockout mice exhibit impaired spatial working memory. *Neuroscience Letters, 668*, 92–97.

de Lecea, L., Kilduff, T. S., Peyron, C., Gao, X., Foye, P. E., Danielson, P. E., et al. (1998). The hypocretins: hypothalamus-specific peptides with neuroexcitatory activity. *Proceedings of the National Academy of Sciences of the United States of America, 95*, 322–327.

Elias, C. F., Saper, C. B., Maratos-Flier, E., Tritos, N. A., Lee, C., Kelly, J., et al. (1998). Chemically defined projections linking the mediobasal hypothalamus and the lateral hypothalamic area. *The Journal of Comparative Neurology*, *402*, 442–459.

Fadel, J., & Deutch, Y. (2002). Anatomical substrates of orexin-dopamine interactions: Lateral hypothalamic projections to the ventral tegmental area. *Neuroscience*, *111*, 379–387.

Fenno, L., Yizhar, O., & Deisseroth, K. (2011). The development and application of optogenetics. *Annual Review of Neuroscience*, *34*, 389–412.

González, J. A., Jensen, L. T., Iordanidou, P., Strom, M., Fugger, L., & Burdakov, D. (2016). Inhibitory interplay between orexin neurons and eating. *Current Biology*, *26*, 2486–2491.

Hahn, J. D., & Swanson, L. W. (2010). Distinct patterns of neuronal inputs and outputs of the juxtaparaventricular and suprafornical regions of the lateral hypothalamic area in the male rat. *Brain Research Reviews*, *64*, 14–103.

Harris, G. C., Wimmer, M., & Aston-Jones, G. (2005). A role for lateral hypothalamic orexin neurons in reward seeking. *Nature*, *437*, 556–559.

Hassani, O. K., Krause, M. R., Mainville, L., Cordova, C. A., & Jones, B. E. (2016). Orexin neurons respond differentially to auditory cues associated with appetitive versus aversive outcomes. *The Journal of Neuroscience*, *36*, 1747–1757.

Haynes, A. C., Jackson, A. C., Overend, P., Buckingham, R. E., Wilson, S., Tadayyon, M., et al. (1999). Effects of single and chronic intracerebroventricular administration of the orexins on feeding in the rat. *Peptides*, *20*, 1099–1105.

Haynes, A. C., Jackson, B., Champan, H., Tadayyon, M., Johns, A., Porter, R. A., et al. (2000). A selective orexin-1 receptor antagonist reduces food consumption in male and female rats. *Regulatory Peptides*, *96*, 45–51.

Ho, C. Y., & Berridge, K. C. (2013). An orexin hotspot in ventral pallidum amplifies hedonic "liking" for sweetness. *Neuropsychopharmacology*, *38*, 1655–1664.

Horvath, T. L., Diano, S., & van den Pol, A. N. (1999). Synaptic intercation between hypocretin (orexin) and neuropeptide Y cells in the rodent and primate hypothalamus: a novel circuit implicated in metabolic and endocrine regulations. *The Journal of Neuroscience*, *19*, 1072–1087.

Hsu, T. M., Konanur, V. R., Noble, E. E., Suarez, A. N., Thai, J., Nakamoto, E. M., et al. (2015). Hippocampus ghrelin signaling mediates appetite through lateral hypothalamic orexin pathways. *eLife*, *4*.

Huang, H., Ghosh, P., & van den Pol, A. N. (2006). Prefrontal cortex–projecting glutamatergic thalamic paraventricular nucleus-excited by hypocretin: A feedforward circuit that may enhance cognitive arousal. *95*, 1656–1668.

Hurley, S. W., & Johnson, A. K. (2014). The role of the lateral hypothalamus and orexin in ingestive behavior: a model for the translation of past experience and sensed deficits into motivated behaviors. *Frontiers in Systems Neuroscience*, *8*, 216.

Jain, M. R., Horvath, T. L., Kalra, P. S., & Kalra, S. P. (2000). Evidence that NPY Y1 receptors are involved in stimulation of feeding by orexins (hypocretins) in sated rats. *Regulatory Peptides*, *87*, 19–24.

James, M. H., Mahler, S. V., Moorman, D. E., & Aston-Jones, G. (2017). A decade of orexin/hypocretin and addiction: Where are we now? *Current Topics in Behavioral Neurosciences*, *33*, 247–281.

Jennings, J. H., Rizzi, G., Stamatakis, A. M., Ung, R. L., & Stuber, G. D. (2013). The inhibitory circuit architecture of the lateral hypothalamus orchestrates feeding. *Science*, *341*, 1517–1521.

Kay, K., Parise, E. M., Lilly, N., & Williams, D. L. (2014). Hindbrain orexin 1 receptors influence palatable food intake, operant responding for food, and food-conditioned place preference in rats. *Psychopharmacology*, *231*, 419–427.

Keefer, S. E., Cole, S., & Petrovich, G. D. (2016). Orexin/hypocretin receptor 1 signaling mediates Pavlovian cue-food conditioning and extinction. *Physiology & Behavior, 162*, 27–36.

Leinninger, G. M., Opland, D. M., Jo, Y. H., Faouzi, M., Christensen, L., Cappellucci, L. A., et al. (2011). Leptin action via neurotensin neurons controls orexin, the mesolimbic dopamine system and energy balance. *Cell Metabolism, 14*, 313–323.

Li, S., & Kirouac, G. J. (2012). Sources of inputs to the anterior and posterior aspects of the paraventricular nucleus of the thalamus. *Brain Structure & Function, 217*, 257–273.

Li, Y., & van den Pol, A. N. (2006). Differential target-dependent actions of coexpressed inhibitory dynorphin and excitatory hypocretin/orexin neuropeptides. *The Journal of Neuroscience, 26*, 13037–13047.

Mahler, S. V., Moorman, D. E., Smith, R. J., James, M. H., & Aston-Jones, G. (2014). Motivational activation: A unifying hypothesis of orexin/hypocretin function. *Nature Neuroscience, 17*, 1298–1303.

Marcus, J. N., Aschkenasi, C. J., Lee, C. E., Chemelli, R. M., Saper, C. B., Yanagisawa, M., et al. (2001). Differential expression of orexin receptors 1 and 2 in the rat brain. *Journal of Comparative Neurology, 435*, 6–25.

Matsuo, E., Mochizuki, A., Nakayama, K., Nakamura, S., Yamamoto, T., Shioda, S., et al. (2011). Decreased intake of sucrose solutions in orexin knockout mice. *Journal of Molecular Neuroscience, 43*, 217–224.

Mavanji, V., Butterick, T. A., Duffy, C. M., Nixon, J. P., Billington, C. J., & Kotz, C. M. (2017). Orexin/hypocretin treatment restores hippocampal-dependent memory in orexin-deficient mice. *Neurobiology of Learning and Memory, 146*, 21–30.

Mediavilla, C., Cabello, V., & Risco, S. (2011). SB-334867-A, a selective orexin-1 receptor antagonist, enhances taste aversion learning and blocks taste preference learning in rats. *Pharmacology, Biochemistry, and Behavior, 98*, 385–391.

Mena, J. D., Selleck, R. A., & Baldo, B. A. (2013). Mu-opioid stimulation in rat prefrontal cortex engages hypothalamic orexin/hypocretin-containing neurons, and reveals dissociable roles of nucleus accumbens and hypothalamus in cortically driven feeding. *The Journal of Neuroscience, 33*, 18540–18552.

Mickelsen, L. E., Kolling, F. W., IV, Chimileski, B. R., Fujita, A., Norris, C., Chen, K., et al. (2017). Neurochemical heterogeneity among lateral hypothalamic hypocretin/orexin and melanin-concentrating hormone neurons identified through single-cell gene expression analysis. *eNeuro, 4*(5). https://doi.org/10.1523/ENEURO.0013-17.2017.

Mieda, M., Williams, S. C., Sinton, C. M., Richardson, J. A., Sakurai, T., & Yanagisawa, M. (2005). Orexin neurons function in an efferent pathway of a food-entrainable circadian oscillator in eliciting food-anticipatory activity and wakefulness. *The Journal of Neuroscience, 24*, 10493–10501.

Moorman, D. E. (2018). The hypocretin/orexin system as a target for excessive motivation in alcohol use disorders. *Psychopharmacology, 235*(6), 1663–1680.

Muschamp, J. W., Hollander, J. A., Thompson, J. L., Voren, G., Hassinger, L. C., Onvani, S., et al. (2014). Hypocretin (orexin) facilitates reward by attenuating the antireward effects of its cotransmitter dynorphin in ventral tegmental area. *Proceedings of the National Academy of Sciences of the United States of America, 111*, E1648–E1655.

Nair, S. G., Golden, S. A., & Shaham, Y. (2008). Differential effects of the hypocretin 1 receptor antagonist SB 334867 on high-fat food self-administration and reinstatement of food seeking in rats. *British Journal of Pharmacology, 154*, 406–416.

Navarro, M., Olney, J. J., Burnham, N. W., Mazzone, C. M., Lowery-Gionta, E. G., Pleil, K. E., et al. (2016). Lateral hypothalamus GABAergic neurons modulate consummatory behaviors regardless of the caloric content or biological relevance of the consumed stimuli. *Neuropsychopharmacology, 41*, 1505–1512.

O'Connor, E., Kremer, Y., Lefort, S., Harada, M., Pascoli, V., Rohner, C., et al. (2015). Accumbal D1R neurons projecting to lateral hypothalamus authorize feeding. *Neuron*, *88*, 553–564.

Parsania, S., Moradi, M., Fatahi, Z., & Haghparast, A. (2016). Involvement of orexin-1 and orexin-2 receptors within the dentate gyrus of the hippocampus in the acquisition, expression and extinction of lateral hypothalamic-induced conditioned place preference in the rats. *1639*, 149–160.

Petrovich, G. D. (2013). Forebrain networks and the control of feeding by environmental learned cues. *Physiology & Behavior*, *121*, 10–18.

Petrovich, G. D. (2018). Lateral hypothalamus as a motivation-cognition interface in the control of feeding behavior. *Frontiers in Systems Neuroscience*, *12*, 14.

Petrovich, G. D., Hobin, M. P., & Reppucci, C. J. (2012). Selective Fos induction in hypothalamic orexin/hypocretin, but not melanin-concentrating hormone neurons, by a learned food-cue that stimulates feeding in sated rats. *Neuroscience*, *224*, 70–80.

Peyron, C., Tighe, D. K., van den Pol, A. N., de Lecea, L., Heller, H. C., Sutcliffe, J. G., et al. (1998). Neurons containing hypocretin (orexin) project to multiple neuronal systems. *The Journal of Neuroscience*, *18*, 9996–10015.

Reppucci, C. J., & Petrovich, G. D. (2016). Organization of connections between the amygdala, medial prefrontal cortex, and lateral hypothalamus: A single and double retrograde tracing study in rats. *Brain Structure & Function*, *221*, 2937–2962.

Rosin, D. I., Weston, M. C., Sevigny, C. P., Stornetta, R. I., & Guyenet, P. G. (2003). Hypothalamic orexin (hypocretin) neurons express vesicular glutamate tansporters VGLUT1 or VGLUT2. *The Journal of Comparative Neurology*, *465*, 593–603.

Sakurai, T. (2014). The role of orexin in motivated behaviours. *Nature Reviews. Neuroscience*, *15*, 719–731.

Sakurai, T., Amemiya, A., Ishii, M., Matsuzaki, I., Chemelli, R. M., Tanaka, H., et al. (1998). Orexins and orexin receptors: a family of hypothalamic neuropeptides and G protein-coupled receptors that regulate feeding behavior. *Cell*, *92*, 573–585.

Sakurai, T., Nagata, R., Yamanaka, A., Kawamura, H., Tsujino, N., Muraki, Y., et al. (2005). Input of orexin/hypocretin neurons revealed by a genetically encoded tracer in mice. *Neuron*, *46*, 297–308.

Saper, C. B. (2006). Staying awake for dinner: hypothalamic integration of sleep, feeding, and circadian rhythms. *Progress in Brain Research*, *153*, 243–252.

Scammell, T., Arrigoni, E., & Lipton, J. O. (2017). Neural circuitry of wakefulness and sleep. *Neuron*, *93*, 747–765.

Scammell, T. E., & Winrow, C. J. (2011). Orexin receptors: Pharmacology and therapeutic opportunities. *Annual Review of Pharmacology and Toxicology*, *51*, 243–266.

Schöne, C., & Burdakov, D. (2012). Glutamate and GABA as rapid effectors of hypothalamic "peptidergic" neurons. *Frontiers in Behavioral Neuroscience*, *6*, 81.

Sharf, R., Sarhan, M., Brayton, C. E., Guarnieri, D. J., Taylor, J. R., & DiLeone, R. J. (2010). Orexin signaling via the orexin 1 receptor mediates operant responding for food reinforcement. *Biological Psychiatry*, *67*, 753–760.

Sharf, R., Sarhan, M., & DiLeone, R. J. (2010). Role of orexin/hypocretin in dependence and addiction. *Brain Research*, *1314*, 130–138.

Stamatakis, A. M., Swieten, M. V., Basiri, M. L., Blair, G. A., Kantak, P., & Stuber, G. D. (2016). Lateral hypothalamic area glutamatergic neurons and their projections to the lateral habenula regulate feeding and reward. *The Journal of Neuroscience*, *36*, 302–311.

Sternson, S. M., & Roth, B. L. (2014). Chemogenetic tools to interrogate brain functions. *Annual Review of Neuroscience*, *37*, 387–407.

Sun, X., Kroemer, N. B., Veldhuizen, M., Babbs, A. E., de Araujo, I. E., Gitelman, D. R., et al. (2015). Basolateral amygdala response to food cues in the absence of hunger is associated with weight gain susceptibility. *The Journal of Neuroscience*, *35*, 7964–7976.

Swanson, L. W. (2000). Cerebral hemisphere regulation of motivated behavior. *Brain Research*, *886*, 113–164.
Swanson, L. W. (2004). *Brain maps: structure of the rat brain: A laboratory guide with printed and electronic templates for data, models and schematics* (3rd revised edition with CD-ROM ed.). Amsterdam: Elsevier.
Swanson, L. W., Sanchez-Watts, G., & Watts, A. G. (2005). Comparison of melanin-concentrating hormone and hypocretin/orexin mRNA expression patterns in a new parceling scheme of the lateral hypothalamic zone. *Neuroscience Letters*, *387*, 80–84.
Sweet, D. C., Levine, A. S., Billington, C. J., & Kotz, C. M. (1999). Feeding response to central orexins. *Brain Research*, *821*, 535–538.
Thorpe, A. J., Cleary, J. P., Levine, A. S., & Kotz, C. M. (2005). Centrally administered orexin A increases motivation for sweet pellets in rats. *Psychopharmacology*, *182*, 75–83.
Thorpe, A. J., & Kotz, C. M. (2005). Orexin A in the nucleus accumbens stimulates feeding and locomotor activity. *Brain Research*, *1050*, 156–162.
Tyree, S. M., & de Lecea, L. (2017). Lateral hypothalamic control of the ventral tegmental area: reward evaluation and the driving of motivated behavior. *Frontiers in Systems Neuroscience*, *11*, 50.
Weingarten, H. P. (1983). Conditioned cues elicit feeding in sated rats: A role for learning in meal initiation. *Science*, *220*, 431–433.
Woods, S. C., & Ramsay, D. S. (2000). Pavlovian influence over food and drug intake. *Behavioural Brain Research*, *110*, 175–182.
Yamanaka, A., Beuckmann, C. T., Willie, J. T., Hara, J., Tsujino, N., Mieda, M., et al. (2003). Hypothalamic orexin neurons regulate arousal according to energy balance in mice. *Neuron*, *38*, 701–713.
Yamanaka, A., Kunii, K., Nambu, T., Tsujino, N., Sakai, A., Matsuzaki, I., et al. (2000). Orexin-induced food intake involves neuropeptide Y pathway. *Brain Research*, *859*, 404–409.
Yoshida, K., McCormack, S., España, R. A., Crocker, A., & Scammell, T. (2006). Afferents to the orexin neurons of the rat brain. *The Journal of Comparative Neurology*, *494*, 845–861.
Zheng, H., Patterson, L. M., & Berthoud, H. R. (2007). Orexin signaling in the ventral tegmental area is required for high-fat appetite induced by opioid stimulation of the nucleus accumbens. *The Journal of Neuroscience*, *27*, 11075–11082.

CHAPTER 5

Hypocretin Receptor 1 Regulation of Dopamine Neurotransmission and Motivated Behavior

Jessica K. Shaw*, Emily M. Black*, Yanan Zhang†, Rodrigo A. España*
*Department of Neurobiology and Anatomy, Drexel University College of Medicine, Philadelphia, PA, United States
†Research Triangle Institute, Research Triangle Park, NC, United States

INTRODUCTION

Hypocretins (HCRT, also known as orexins) are peptides synthesized by neurons originating in the lateral hypothalamus and adjacent perifornical areas. Neurons that produce HCRT project widely throughout the brain, producing behavioral effects via actions at the HCRT receptor 1 and HCRT receptor 2 (HCRTr1, HCRTr2; de Lecea et al., 1998; Fadel & Deutch, 2002; Sakurai et al., 1998). Although originally posited to elicit feeding and later found to govern arousal-related processes, mounting evidence now implicates these peptides and their receptors in the regulation of reward and reinforcement (Boutrel et al., 2005; Harris, Wimmer, & Aston-Jones, 2005). In the following sections, we posit that HCRT exerts its effects through actions on the mesolimbic dopamine (DA) system. Indeed, the ventral tegmental area (VTA)—alongside the adjacent dopaminergic substantia nigra pars compacta (SN)—strongly expresses both HCRTr1 and HCRTr2 (Marcus et al., 2001), although VTA HCRTr1 appears to be the primary mediator of motivated behavior (Prince, Rau, Yorgason, & España, 2015). These observations, coupled with multiple clinical studies (Khoo & Brown, 2014), have led to the consideration of HCRT-based pharmacotherapies for treating addiction.

HCRTr1 SUPPORTS DRUG-ASSOCIATED BEHAVIOR

Addiction is a complex disorder characterized by heavy drug consumption, excessive motivation to obtain drug, and relapse to drug-seeking following

The Orexin/Hypocretin System
https://doi.org/10.1016/B978-0-12-813751-2.00005-X

drug-free periods. The impact of HCRTr1 on each of these behaviors varies according to the drug class examined. In this section, we provide a brief overview of the influence of HCRTr1 activity on major drug-associated behaviors.

HCRTr1 Modulates Conditioned Place Preference

Conditioned place preference (CPP) is among the most widely used behavioral assays in the study of addiction (Tzschentke, 2007). This paradigm involves administering a drug of interest while confining the animal to a specific chamber for one session and administering a vehicle before confining the animal to another chamber in a separate session. Each pairing is repeated multiple times to develop a Pavlovian association between the drug and its paired chamber. Following the conditioning period, a probe session is conducted during which animals are allowed to freely explore the behavioral apparatus (Bardo & Bevins, 2000; Tzschentke, 1998, 2007). CPP is demonstrated when the time spent in the drug-paired chamber is greater than the time spent in the vehicle-paired chamber. While CPP has occasionally been criticized for its lack of face validity, the vast majority of addictive substances produce CPP and it is frequently used to screen substances for abuse liability (Bardo & Bevins, 2000; Tzschentke, 1998).

The contribution of HCRT neurotransmission to drug-associated behaviors has been well studied using this assay. HCRT knock out (KO) mice fail to demonstrate CPP for cocaine (Shaw et al., 2016), an effect that appears to be mediated by HCRTr1, as the HCRTr1 antagonist SB-334867 inhibits the expression of CPP for both cocaine (Sartor & Aston-Jones, 2012) and amphetamine (Hutcheson et al., 2011). Similarly, HCRTr1 blockade prevents both the development (acquisition) and expression of morphine-induced place preference (Farahimanesh, Zarrabian, & Haghparast, 2017). Contrary to observations using psychostimulants in HCRT KO mice, HCRTr1 antagonism fails to inhibit the development of place preference for alcohol, although it reduces the expression of CPP in animals that had not been habituated to injections (Voorhees & Cunningham, 2011). Importantly, HCRT KO mice maintain the ability to form conditioned place aversion to lithium chloride (Narita et al., 2006), suggesting that observed alterations in CPP are not due to an inability to form associative memories.

HCRTr1 Supports High-Effort Drug Self-Administration

Significant progress toward understanding HCRTr1's function in drug self-administration has been made using cocaine as a reinforcer. Initially, the effects of HCRT agents were tested on cocaine self-administration under the fixed-ratio (FR) 1 schedule of reinforcement. This widely used, low-effort schedule requires the completion of a single operant response to obtain a single reinforcer. Under this schedule, rats treated with HCRT-1 peptide or an HCRTr1 antagonist fail to alter their consumption of a moderate dose of cocaine (España et al., 2010; España, Melchior, Roberts, & Jones, 2011). These findings were initially interpreted to represent a failure of HCRT compounds to alter ongoing self-administration. However, as detailed later, it is now evident that HCRT-based manipulations preferentially affect high-effort responding for cocaine.

A measure of motivation may be obtained by manipulating the effort to reward ratio, or unit price, across a self-administration session. The progressive ratio (PR) schedule increases the number of operant responses required to obtain each reinforcer across the session, thus requiring increasingly more effort to maintain preferred blood levels of cocaine as the session progresses. Under this schedule, HCRTr1 antagonism significantly decreases effortful responding for cocaine (Borgland et al., 2009; Brodnik, Bernstein, Prince, & España, 2015; España et al., 2010; Levy et al., 2017; Prince et al., 2015). Importantly, HCRTr2 antagonism fails to reduce motivation under this schedule, indicating the specificity of HCRTr1's involvement in this behavior (Fig. 1; Prince et al., 2015).

Manipulating the unit price by changing the dose of the cocaine reinforcer produces similar results when HCRTr1 is disrupted. Although at moderate doses of cocaine (0.75 mg/kg/injection) HCRTr1 disruption fails to alter low-effort responding (Brodnik et al., 2015; España et al., 2010), HCRTr1 blockade successfully reduces low-effort responding when the dose of the cocaine is decreased (0.375 mg/kg/injection). Conversely, increasing the dose of cocaine under the PR schedule—i.e., decreasing the unit price—diminishes the effect of HCRTr1 antagonism on responding (Brodnik et al., 2015).

The within-session threshold procedure (Threshold) also manipulates unit price over the course of a single session. Contrary to PR, Threshold manipulates the effort to reward ratio by maintaining the effort requirement and deescalating the reinforcer value in a stepwise fashion. This assay permits a measure of "demand" early in the session when the drug is essentially free, as well as a measure of motivation represented by the maximal unit price that maintains responding (Pmax; Oleson & Roberts, 2012). Similar to results using FR1, HCRTr1 antagonism did not affect demand for cocaine under the Threshold schedule. Nevertheless, HCRTr1 antagonism significantly

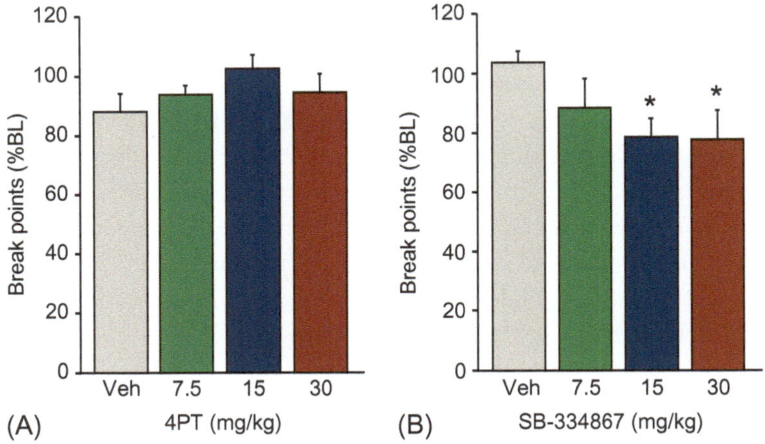

Fig. 1 Blockade of HCRTr1, but not HCRTr2, reduces motivation for cocaine. Shown are mean ± SEM break points expressed as a percent of baseline in animals treated with the (A) HCRTr2 antagonist 4PT or the (B) HCRTr1 antagonist SB-334867 during a progressive ratio session. While HCRTr2 antagonism failed to affect responding, HCRTr1 significantly reduced break points. *$P < .05$. *(Modified from Prince, C.D., Rau, A.R., Yorgason, J.T., & España, R.A. (2015). Hypocretin/orexin regulation of dopamine signaling and cocaine self-administration is mediated predominantly by hypocretin receptor 1.* ACS Chemical Neuroscience, *6(1), 138–146. https://doi.org/10.1021/cn500246j.)*

reduces Pmax (España et al., 2010). Together with the PR findings, these results suggest that HCRTr1 impacts motivation for cocaine by supporting drug-directed behavior in situations when the unit price is high.

In addition to classical measures of motivation, the contribution of HCRTr1 signaling to drug-seeking behavior has been investigated using the otherwise underutilized discrete trials (DT) task. Although this schedule is based on the low-effort FR1, animals are limited in the number of infusions they can self-administer each hour over a 24-hr period and thus are unable to titrate to preferred blood levels. This schedule yields a measure of the animal's willingness to subvert the physiological drive for sleep, as rodents respond primarily during the dark/active phase of their light cycle and respond only minimally during the light phase, when they are typically at rest (Roberts, Brebner, Vincler, & Lynch, 2002). Rats responding under this schedule demonstrated reduced cocaine intake during the dark cycle when HCRTr1 was disrupted, which was associated with early cessation of cocaine-taking during the session (España et al., 2010), and increased responding when treated with the HCRT-1 peptide, which effectively prolonged cocaine-taking into the light phase (España et al., 2011).

Self-administration paradigms rely on the exertion of physical effort to obtain a reinforcer, thus raising concerns that differences in operant responding may be associated with alterations in arousal and/or motor activity rather than specific actions on motivation. This is particularly true with manipulations of HCRT, given the extensive literature implicating HCRT in arousal and sleep/wake processes. Importantly, neither acute (Brodnik et al., 2015) nor sustained (Bernstein, Badve, Barson, Bass, & España, 2017) disruption of HCRTr1 signaling induces sleep or sedation, nor does HCRTr1 blockade appear to affect locomotor activity, particularly in response to cocaine. As depicted in Fig. 2, rats were treated with the highly selective HCRTr1 antagonist RTIOX-276 at varying doses 30 min prior to being placed in a novel locomotor chamber. Locomotor activity was measured by automated calculation of photobeam breaks and represented as distance traveled (m). During this initial test, we observed no effects of HCRTr1 antagonism on locomotor activity [Treatment: $F_{(3,20)} = 0.994$, $P = .416$; Time × Treatment: $F_{(21,140)} = 1.397$, $P = .129$]. After 40 min of testing, rats were treated with 10 mg/kg cocaine intraperitoneally (i.p.) and activity was recorded for another 90 min. Consistent with previous work using another HCRTr1 antagonist (Calipari & España, 2012), we observed that blocking HCRTr1 did not affect cocaine-induced increases

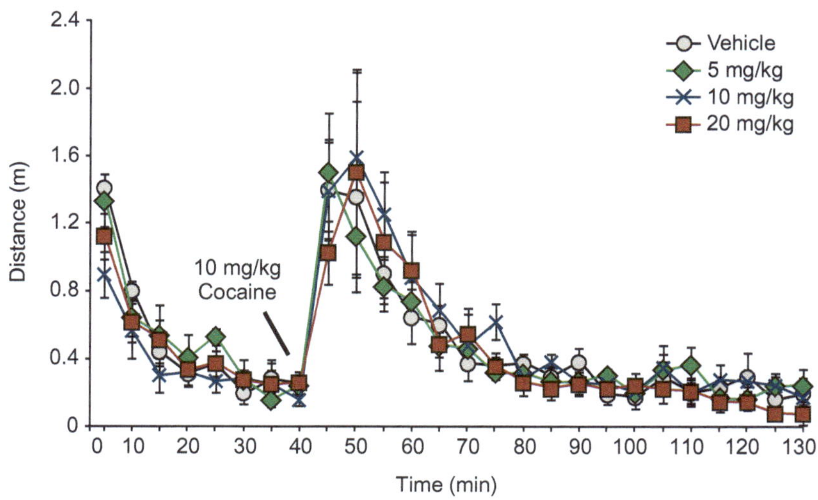

Fig. 2 HCRTr1 antagonism does not reduce cocaine-induced locomotor activity. Shown are the mean ± SEM of locomotor activity expressed as distance traveled (m) for animals treated with i.p. vehicle ($n = 6$) or the highly selective HCRTr1 antagonist RTIOX-276 (5, 10, or 20 mg/kg $n = 6$) before and after administration of cocaine (10 mg/kg, i.p.). HCRTr1 antagonism did not significantly alter locomotor activity relative to vehicle treatment.

in locomotor activity [Treatment: $F_{(3,20)} = 0.150$, $P = .0928$; Time × Treatment: $F_{(51,140)} = 0.421$, $P = 1.0$]. This failure of HCRTr1 blockade to either diminish cocaine-induced locomotor activity or to alter sleep strongly suggests that the effects of HCRTr1 on drug-associated behavior are unlikely to be the result of gross alterations in arousal or deficits in motor activity.

The effects of HCRT manipulations on opioid self-administration have not been as extensively studied as psychostimulants. Unlike psychostimulants, both HCRTr1 and HCRTr2 disruption reduce heroin self-administration under low- (Schmeichel et al., 2015; Smith & Aston-Jones, 2012) and high-effort schedules (Smith & Aston-Jones, 2012). Intriguingly, HCRTr1 disruption fails to significantly alter FR1 responding for remifentanil—a potent, ultra-short acting opioid—suggesting that the pharmacodynamics of the drug reinforcer may be important to the efficacy of HCRTr1-based interventions. Additionally, while HCRTr1 disruption reduces motivation for remifentanil under the Threshold schedule it also decreases demand, suggesting that HCRTr1 signaling mediates the reinforcing properties of opioids in a way it does not for cocaine.

Similar to opioids, HCRTr1 antagonism decreases responding for nicotine under both relatively low (FR5) and high (PR) effort schedules of reinforcement (Hollander, Lu, Cameron, Kamenecka, & Kenny, 2008; LeSage, Perry, Kotz, Shelley, & Corrigall, 2010). The effect of HCRTr1 activation on alcohol self-administration, however, is less clear. Nevertheless, blockade of HCRTr1 has been shown to reduce FR3 responding for alcohol (Jupp, Krivdic, Krstew, & Lawrence, 2011) and decrease alcohol self-administration under the PR schedule (Jupp, Krivdic, et al., 2011). Together, these data indicate that HCRTr1 signaling supports high-effort responding for all substances of abuse but may have a more complex effect profile in consumption depending on the drug in question.

HCRTr1 Supports Relapse/Reinstatement

Addiction is frequently characterized as a chronic relapsing disorder, and recovery is hampered by relapse to drug use. Relapse is often modeled using reinstatement procedures, in which animals are trained to self-administer drugs and then allowed to extinguish the relationship between the operant response and the reinforcer. They are then challenged with a stressor, a priming injection of the drug, or with a drug-paired cue and the resulting operant responses are measured.

HCRTr1 Supports Cocaine Reinstatement

Broadly speaking, HCRTr1 supports reinstatement of cocaine self-administration, although its impact appears to be dependent on the specific triggers used to reinstate the conditioned behavior. When cocaine-seeking is reinstated using discrete, discriminative, or contextual cues, HCRTr1 antagonism reduces cocaine-seeking compared to vehicle (Bentzley & Aston-Jones, 2015; James et al., 2011; Smith, Tahsili-Fahadan, & Aston-Jones, 2010; Zhou et al., 2012)—suggesting that signaling at HCRTr1 supports relapse to drug-taking in response to drug-associated cues (Smith et al., 2010).

Stress is another major factor implicated in relapse to drug use and HCRTr1 signaling is believed to be a key player in this effect. In mice, cocaine CPP was reinstated by acute restraint stress, which was prevented by systemic and intra-VTA HCRTr1 antagonism (Tung et al., 2016). Similarly, yohimbine—a pharmacological stressor—robustly reinstates cocaine-associated behavior under normal conditions, and this reinstatement phenomenon is attenuated following HCRTr1 antagonism (Zhou et al., 2012). Interestingly, although peripheral administration of an HCRTr1 antagonist inhibits footshock-induced reinstatement (Boutrel et al., 2005), intra-VTA treatment with the HCRTr1 antagonist SB-408124 fails to elicit the same effect (Wang, You, & Wise, 2009). Consistent with its lack of effect on the reinforcing effects of cocaine, HCRTr1 does not appear to support reinstatement in response to a priming injection of psychostimulants; rats pretreated with an HCRTr1 antagonist maintain robust reinstatement following a cocaine injection (Zhou et al., 2012).

HCRTr1 Supports Cue-Induced Reinstatement of Other Drugs of Abuse

Similar to psychostimulants, HCRTr1 signaling supports reinstatement of opioid- and ethanol-seeking behavior. HCRTr1 blockade dose dependently reduces cue-induced reinstatement for heroin and remifentanil but fails to significantly affect reinstatement following a drug-priming injection (Porter-Stransky, Bentzley, & Aston-Jones, 2017; Smith & Aston-Jones, 2012). Similarly, HCRTr1 antagonism decreases cue-evoked reinstatement of ethanol-seeking after both short and long periods of imposed abstinence in alcohol-preferring rats (Jupp, Krstew, Dezsi, & Lawrence, 2011). HCRTr1 antagonism also inhibits stress-induced reinstatement of ethanol-seeking without affecting sucrose-seeking (Richards et al., 2008). As with opioids and psychostimulants, HCRTr1 antagonists decrease cue-induced reinstatement of nicotine-seeking behavior. Mice trained to self-administer i.v. nicotine under FR1 show significant reductions in cue-induced reinstatement following treatment with an HCRTr1—but not HCRTr2—antagonist, suggesting that actions at HCRTr1 support cue-induced reinstatement of nicotine self-administration in addition to opioids and psychostimulants (Plaza-Zabala et al., 2013). Unlike cocaine and ethanol, however, HCRTr1 does not appear to influence stress-induced reinstatement of nicotine-seeking. Although both footshock and intracerebral ventricular (i.c.v.) administration of HCRT-1 induce reinstatement of nicotine-seeking, HCRTr1 blockade fails to prevent footshock-induced reinstatement on its own, though it robustly inhibits the reinstatement effects of HCRT-1 (Plaza-Zabala, Martin-Garcia, de Lecea, Maldonado, & Berrendero, 2010).

HCRTr1 and Motivation for Food Rewards

Drugs of abuse are believed to have co-opted some of the neurocircuitry underlying natural reinforcement (Kelley & Berridge, 2002), thus it is not surprising that HCRTr1 supports motivation for nonhomeostatic food consumption, particularly of highly rewarding food. For example, HCRTr1 blockade reduces consumption of high-fat—but not nutritionally balanced—chow (Borgland et al., 2009). Similarly, HCRTr1 antagonism reduces the effort expended to obtain sucrose, but the magnitude of this effort inhibition depends on the satiety state—and, thus, caloric deficit—of the animal. HCRTr1 blockade significantly inhibits motivation to obtain sucrose pellets in sated rats, whereas it fails to decrease responding in fasted (i.e., hungry) animals (Cason & Aston-Jones, 2013; España et al., 2010). Similar to drugs of abuse, HCRTr1 appears to preferentially support some forms of reinstatement of palatable food-seeking more than others. For example, HCRTr1 blockade inhibits cue-induced reinstatement of sucrose-seeking (Cason & Aston-Jones, 2013) but fails to attenuate pellet-induced (analogous to drug-primed) reinstatement (Nair, Golden, & Shaham, 2008). The effect of HCRTr1 blockade on stress-induced reinstatement of sucrose-seeking is less clear, with studies showing both no effect (Nair et al., 2008) and a significant reduction (Richards et al., 2008) of stress-induced reinstatement. Together, these findings suggest that while HCRTr1 signaling does not mediate the direct physiological drive for food, it may nevertheless support the motivational drive for nonhomeostatic feeding.

HCRTr1 AND THE MESOLIMBIC DA SYSTEM

Numerous neural pathways have been implicated in cocaine use disorders; however, the mesolimbic DA projections from the VTA to the nucleus accumbens (NAc; Ikemoto, 2007) have garnered the most intense study. Increased extracellular DA is a hallmark response to the majority of reinforcing stimuli—including addictive drugs, food, and sex—and activity in the NAc, in particular, has been associated with motivated behaviors (Spanagel & Weiss, 1999). While the specific function of DA neurotransmission in drug abuse, reward, and addiction liability is a matter of some dispute (Berridge, 2012; Salamone, Correa, Mingote, & Weber, 2005), it has nevertheless been consistently demonstrated to mediate the reinforcing effects of various drugs of abuse (Rassnick, Pulvirenti, & Koob, 1992; Roberts, Corcoran, & Fibiger, 1977). Indeed, the majority of drugs of abuse—including cocaine, ethanol, nicotine, and opioids—increase DA release in the NAc (Burton, Nakamura, & Roesch, 2015; Carboni, Imperato, Perezzani, & Di Chiara, 1989; Di Chiara & Imperato, 1988), albeit through distinct mechanisms.

HCRTr1 Supports DA Neurotransmission Under Drug-Free Conditions

By virtue of its positioning within the mesolimbic DA system, HCRTr1 stands as a promising player in substance abuse and the treatment for addiction. DA neurons of the VTA express both HCRTr1 and HCRTr2 (Fadel & Deutch, 2002), and HCRTs increase intracellular calcium (Narita et al., 2007), induce burst firing, and potentiate glutamatergic excitation of DA neurons (Borgland, Taha, Sarti, Fields, & Bonci, 2006; Korotkova, Sergeeva, Eriksson, Haas, & Brown, 2003), suggesting that HCRT acts in the VTA to support DA release. Early evidence for this relationship came from experiments demonstrating that intra-VTA infusions of the HCRT-1 peptide increase extracellular DA in the prefrontal cortex (Vittoz & Berridge, 2006) and NAc (Narita et al., 2006, 2007). Conversely, intra-VTA HCRTr1 blockade reduces both extracellular concentrations and evoked release of DA (España et al., 2010). Contrary to acute, intra-VTA pharmacological blockade of HCRTr1 in rats (España et al., 2010), HCRT KO mice show no difference in basal extracellular DA relative to WT mice (Narita et al., 2006; Shaw et al., 2016). Nevertheless, when evoked DA release is measured using fast scan cyclic voltammetry (voltammetry), HCRT KO mice consistently show reduced DA release in the NAc ex vivo (España et al., 2010) and in vivo (Shaw et al., 2016). Likewise, rats with genetic knockdown of HCRTr1 in the VTA showed similar results to these HCRT KO mice, demonstrating reduced evoked DA release relative to controls (Bernstein et al., 2017). Intriguingly, disruption of HCRTr1 either genetically (Bernstein et al., 2017; Shaw et al., 2016) or pharmacologically (Levy et al., 2017) decreases the rate of DA uptake (*V*max) in addition to reducing DA release, although the mechanisms underlying this effect remain unclear.

HCRTr1 Supports DA Responses to Various Substances of Abuse

Virtually all drugs of abuse increase extracellular concentrations of DA in the NAc (Di Chiara & Imperato, 1988). As mentioned previously, DA neurons of the VTA express HCRTr1 and are excited by HCRT peptides (Korotkova et al., 2003). This relationship extends to DA neurotransmission in response to drugs of abuse. Notably, the majority of work studying the interaction of HCRTr1 signaling and DA has focused on cocaine. In the following section, we survey some of the major findings.

HCRTr1 and DA Responses to Drugs of Abuse

Cocaine increases DA in the NAc by blocking the DA transporter (DAT), thus inhibiting DA uptake and prolonging the presence of DA extracellularly. Cocaine-induced changes in DA dynamics are readily detectable using multiple techniques, rendering it an ideal model drug to investigate the contribution of HCRTr1 to DA neurotransmission. HCRTr1 activity supports robust DA responses to cocaine. HCRTr1 antagonists—but not HCRTr2 antagonists—delivered i.p. prior to cocaine delivery inhibit cocaine-induced DA uptake inhibition, thus demonstrating that HCRTr1 mediates cocaine-induced changes in DA signaling (Levy et al., 2017; Prince et al., 2015). Moreover, HCRTr1 blockade in the VTA produces similar reductions in cocaine-induced uptake inhibition. Interestingly, in contrast to the effects in rats treated systemically with HCRTr1 antagonists, intra-VTA infusions also produce significant reductions in the peak height of evoked DA release (España et al., 2010). Additionally, HCRTr1 blockade decreases the amplitude of spontaneous DA release events under both drug-free conditions and in response to cocaine (Levy et al., 2017; see later).

The effects of cocaine in animals with genetically modified HCRT neurotransmission are largely consistent with effects seen following acute pharmacological manipulations of HCRTr1. For example, although the basal concentration of DA in the NAc does not differ between HCRT KO and WT mice under drug-free conditions (Shaw et al., 2016), increases in extracellular DA following cocaine are significantly lower in KO mice relative to WT mice. The effect of HCRT deficiency on the DA response to cocaine is similarly reflected in differences in evoked DA as measured using in vivo voltammetry. The reduction in evoked DA release observed under baseline conditions persists following an i.p. cocaine injection in both anesthetized and freely moving HCRT KO mice (Shaw et al., 2016; see Fig. 4), although the apparent ability of cocaine to inhibit the DAT is not different between genotypes. These results are consistent with those from rats with genetic HCRTr1 knockdown in the VTA. As mentioned previously, HCRTr1 knockdown reduces baseline DA release, and this decrease in DA persists following i.v. cocaine delivery. In keeping with acute pharmacological blockade of HCRTr1—but counter to HCRT KO mice—these rats also show a concomitant reduction in cocaine-induced uptake inhibition (Bernstein et al., 2017). Although the mechanisms underlying this difference are unclear, it is possible that congenital HCRT peptide deficiency results in compensatory alterations in DA systems that are not encountered with a relatively brief knockdown of HCRTr1.

The impact of HCRTr1 activity on DA in response to substances of abuse is not limited to drugs with direct actions on DA neurons. Unlike cocaine, opioids are believed to induce increases in extracellular DA indirectly via disinhibition of VTA DA neurons, although numerous observations suggest that opioid actions on mesolimbic DA are more complex (Margolis, Toy, Himmels, Morales, & Fields, 2012). Similar to cocaine, a morphine challenge increases extracellular DA in the NAc in WT mice. However, this morphine-induced increase in DA is significantly inhibited in HCRT KO mice (Narita et al., 2006).

HCRTr1 MODULATION OF DA SYSTEMS SUPPORTS ADDICTION-LIKE BEHAVIORS

Some of the most compelling evidence that HCRTr1 activity modulates drug-associated behavior through actions on the mesolimbic DA system comes from direct manipulations of HCRTr1 in the VTA. Indeed, intra-VTA infusions of HCRT-1 peptide alone are sufficient to produce place preference in a CPP task (Narita et al., 2007), although intra-VTA infusions of HCRT receptor antagonists do not produce any effect (Narita et al., 2006). Further, intra-VTA HCRTr1 blockade reduces locomotor sensitization to cocaine (Borgland et al., 2006) and the motivation for cocaine under a PR schedule of reinforcement (España et al., 2010), implicating the VTA in HCRTr1 regulation of addiction-related behaviors.

Additionally, HCRTr1 in the VTA is important to support the mesolimbic DA response to drug-paired cues. These cues elicit DA release in the NAc where they are believed to underlie drug-context associations that impact cue-induced reinstatement (Covey, Roitman, & Garris, 2014). Indeed, intra-VTA HCRTr1 antagonism reduces cue-induced reinstatement of cocaine-seeking in rats without affecting locomotor activity (James et al., 2011). Further, evidence demonstrates that HCRTr1 blockade inhibits cue-induced DA release and hence impacts cue-induced reinstatement. As shown in Fig. 3, in rats trained to associate cues with noncontingent cocaine injections, blockade of HCRTr1 reduces the amplitude of DA transient release in response to cues, demonstrating that HCRTr1 acts to support cue-induced changes in DA (Levy et al., 2017).

This relationship between HCRTr1 signaling and drug-associated cues persists across pharmacological classes. HCRT-1 into the VTA—but not surrounding areas—is sufficient to reinstate morphine CPP in mice that had previously extinguished this behavior (Harris et al., 2005). Conversely, rats that received intra-VTA—but not intra-SN—infusions of an HCRTr1 antagonist prior to morphine conditioning sessions show dose-dependent reductions in the magnitude of morphine CPP (Narita et al., 2006). Additionally, in line with effects observed in cocaine self-administering animals, intra-VTA HCRTr1 antagonism inhibits cue-induced reinstatement of ethanol-seeking in ethanol-preferring rats (Brown et al., 2016). This effect is specific to the VTA, as rats that received infusions in adjacent areas showed robust reinstatement with no inhibitory effect on ethanol-seeking behavior. Importantly, and in contrast to cocaine, alcohol self-administration has been demonstrated to be altered by both HCRTr1 and HCRTr2 blockade (Brown et al., 2016; Brown, Khoo, & Lawrence, 2013), although cue-induced reinstatement of alcohol-seeking is unaffected by HCRTr2 antagonists.

The persistent, cross-class importance of HCRTr1 actions on the mesolimbic DA system in cue-induced reinstatement may involve interactions between HCRTr1 and glutamatergic AMPA receptors in the VTA. In addition to increasing DA, infusions of intra-VTA HCRT-1, but not HCRT-2, result in increased glutamate release in the VTA and induce reinstatement of cocaine-seeking (Wang et al., 2009). In an elegant study by Mahler et al. (2013), bilateral blockade of either HCRTr1 or glutamatergic receptors

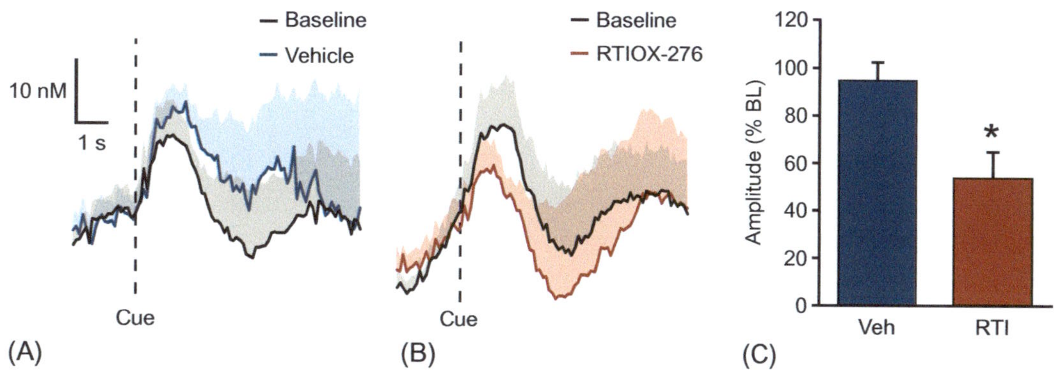

Fig. 3 HCRTr1 blockade inhibits cue-induced phasic DA events. Shown are the mean ± SEM amplitude of cue-evoked DA transients in rats relative to baseline *(black trace)* following treatment with (A) vehicle *(blue trace)* or (B) the HCRTr1 antagonist RTIOX-276 *(red trace)*. *Shaded regions* indicate SEM. HCRTr1 antagonism significantly reduced the amplitude of cue-evoked DA transients as shown in (C; mean ± SEM). *$P < .05$. *(Modified from Levy, K.A., Brodnik, Z.D., Shaw, J.K., Perrey, D.A., Zhang, Y., & España, R.A. (2017). Hypocretin receptor 1 blockade produces bimodal modulation of cocaine-associated mesolimbic dopamine signaling.* Psychopharmacology, *234(18), 2761–2776. https://doi.org/10.1007/s00213-017-4673-y.)*

in the VTA inhibits cue-induced reinstatement of cocaine-seeking but not reinstatement induced by a priming injection of cocaine, consistent with studies of systemic HCRTr1 blockade's effects on reinstatement. This effect persists when an HCRTr1 antagonist was injected into one hemisphere of VTA and AMPA/NMDA antagonists were injected into the contralateral VTA, suggesting that HCRT and glutamate signaling, together, mediate cue-induced reinstatement of cocaine-seeking (Mahler, Smith, & Aston-Jones, 2013).

Additional work suggests that HCRTr1 may support the general responsivity to cues. Rats that self-administered cocaine in the presence of response-contingent cues under the Threshold schedule show significant increases in the elasticity of the demand curve when pretreated with an HCRTr1 antagonist, indicating greater behavioral sensitivity to increases in unit price. By comparison, HCRTr1 blockade had no effect on demand elasticity when responding for cocaine in the absence of cocaine-associated cues (Bentzley & Aston-Jones, 2015). Together with our previous observation that blocking HCRTr1 reduces DA transient responses to cocaine-associated cues (Levy et al., 2017), these findings suggest that HCRTr1 activity is necessary to support the DA response to drug-paired cues and, in turn, the influence of those cues on the risk for addiction.

HCRTr1, DA, AND MOTIVATION IN FEMALES

The majority of work examining HCRT involvement in motivated behavior has focused exclusively on male animals. However, although men are more likely to abuse cocaine (SAMHSA, 2014), women progress to dependence more quickly and show greater severity of abuse-related symptoms (Greenfield, Back, Lawson, & Brady, 2010). This relationship is recapitulated in preclinical studies, which demonstrate that female rats consume higher doses of, work harder for, and are more likely to reinstate cocaine self-administration than male rats (Lynch & Carroll, 2000; Roberts, Bennett, & Vickers, 1989). Recent studies have provided evidence of sex differences in HCRT function and self-administration.

Whereas HCRTr1 blockade decreases motivation for ethanol in alcohol-preferring male rats (Jupp, Krivdic, et al., 2011) it fails to elicit the same effect in females (Anderson, Becker, Adams, Jesudason, & Rorick-Kehn, 2014). Additionally, although HCRTr1 antagonism in female alcohol-preferring rats reduces ethanol consumption following extinction and abstinence, it fails to reduce context-induced reinstatement of ethanol-seeking (Dhaher et al., 2010). In both sexes, the effects of priming injections of cocaine on reinstatement are unaffected by HCRTr1 antagonism, while HCRTr1 blockade inhibits yohimbine-induced reinstatement — although in females this latter effect was only seen at high antagonist doses (Zhou et al., 2012). Contrary to males, however, in which HCRTr1 blockade consistently attenuates cue-induced reinstatement, HCRTr1 antagonism in females fails to affect cue-induced reinstatement of cocaine (Zhou et al., 2012) and sucrose (Cason & Aston-Jones, 2014) seeking. Interestingly, HCRTr1 antagonism reduces ongoing, low-effort sucrose self-administration in females, but achieving this effect required higher doses of the antagonist compared to previous studies in males from the same lab. In contrast, HCRTr1 blockade fails to affect responding for sucrose under PR conditions in these females, although the low response rates in these studies may suggest that rats were not reaching effort levels shown to be sensitive to HCRT manipulations (Brodnik et al., 2015; Cason & Aston-Jones, 2014).

These sex differences in the response to HCRTr1 manipulations may be the result of unique HCRT-DA interactions in males versus females. In a set of experiments designed to examine the influence of HCRT on DA signaling, we evoked DA release in anesthetized WT and HCRT KO male and female mice via electrical stimulation of the VTA and recorded DA release using voltammetry in the NAc. Stimulated release was measured every 5 min until peak heights were stable (see Shaw et al., 2016). Following establishment of baseline DA release, cocaine (10 mg/kg) was administered i.p. and subsequent changes in DA signaling were recorded. As shown in Fig. 4, although male HCRT KO mice display significantly reduced DA at baseline compared to WT mice—an effect that persists following cocaine (Shaw et al., 2016)—female HCRT KO mice do not significantly differ from female WT mice [one-way ANOVA: $F_{(1,15)}=0.345$, $P=.566$] and even show a tendency for slightly higher DA release, although this effect does not reach significance [Genotype: $F_{(1,15)}=0.231$, $P=.638$; Genotype × Time: $F_{(2.302,\ 34.523)}=0.939$, $P=.412$]. Additionally, although male HCRT KO mice demonstrate significantly slower rates of DA uptake compared to WT (Shaw et al., 2016) the rate of DA uptake does not significantly differ between KO and WT female mice (data not shown). These results suggest that HCRT neurotransmission may not exert the same modulatory effects on DA in females as in males, although the mechanisms underlying these differences remain unknown.

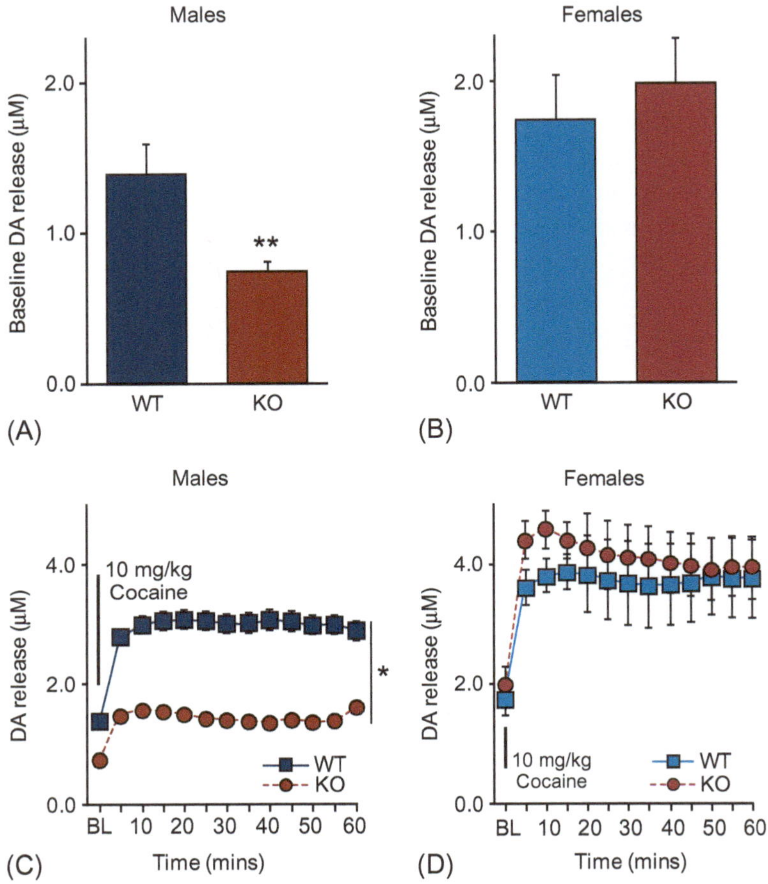

Fig. 4 Male, but not female, HCRT KO mice show reductions in DA at baseline and after cocaine. Shown is mean ± SEM DA release in male and female wild-type (WT) and HCRT knock out (KO) mice. (A) Male HCRT KO mice ($n=7$) have significantly reduced DA release under baseline conditions compared to WT mice ($n=8$; Shaw et al., 2016). However, (B) female WT ($n=9$) and HCRT KO ($n=8$) mice show no significant difference in baseline levels of DA release. This relationship persists following cocaine, in which (C) male HCRT KO mice continue to demonstrate significantly reduced levels of DA compared to WT while (D) female HCRT KO mice do not significantly differ from WT. $^{**}P<.01$; $^{*}P<.05$. *(Male data modified from Shaw, J.K., Ferris, M.J., Locke, J.L., Brodnik, Z.D., Jones, S.R., & España R.A. (2016). Hypocretin/orexin knock-out mice display disrupted behavioral and dopamine responses to cocaine.* Addiction Biology, *22(6), 1695–1705. https://doi.org/10.1111/adb.12432.)*

These results bear important implications for our understanding of the contribution of HCRTr1 to motivation and cue-induced reinstatement. Males and females do not differ in extinction responding, and both male and female rats demonstrate cue-induced reinstatement (Fuchs, Evans, Mehta,

Case, & See, 2005; Zhou et al., 2012). However, previous work suggests that females attribute less motivational salience to cocaine-paired cues relative to males (Fuchs et al., 2005). This suggests that while HCRTr1 signaling renders males vulnerable to cue-induced reinstatement of drug-seeking by supporting DA release in the mesolimbic DA system it fails to produce the same effects in females. These observations may have significant implications for the efficacy of HCRTr1-based pharmacotherapeutic interventions.

CONCLUSIONS AND FUTURE DIRECTIONS

Since its discovery in 1998, the HCRT system has been identified as an important neuromodulator of behavior. By virtue of their expansive projections, HCRT neurons are poised to modulate numerous neurotransmitter systems and thereby impact an array of behaviors. Direct connections with the mesolimbic pathway implicate this system in motivated behavior for both natural and drug rewards, likely through HCRTr1 activity on DA neurons of the VTA. Given that DA transient release in response to drug-associated cues is hypothesized to underlie drug-seeking, we posit that the effects of HCRT manipulations on behavior are mediated by disrupted DA release in response to drug-associated cues.

Although our understanding of HCRT's contribution to addiction-related processes has expanded considerably, much remains unknown regarding the mechanisms underlying HCRT influences on behavioral responses to drug-paired cues and their effect on motivated behavior. Future work examining HCRT-DA interactions in females and males will provide fertile ground to explore these relationships. We believe that understanding the factors that influence DA release in response to drug-associated cues will offer a deeper appreciation of the addiction process, while providing a possible pharmacotherapeutic target to treat the risk of relapse in recovering addicts.

FUNDING

This work was funded by NIDA grant DA031900 and an administrative supplement to support the inclusion of females (DA031900-04S1).

REFERENCES

Anderson, R. I., Becker, H. C., Adams, B. L., Jesudason, C. D., & Rorick-Kehn, L. M. (2014). Orexin-1 and orexin-2 receptor antagonists reduce ethanol self-administration in high-drinking rodent models. *Frontiers in Neuroscience*, *8*, 33. https://doi.org/10.3389/fnins.2014.00033.

Bardo, M. T., & Bevins, R. A. (2000). Conditioned place preference: What does it add to our preclinical understanding of drug reward? *Psychopharmacology*, *153*(1), 31–43.

Bentzley, B. S., & Aston-Jones, G. (2015). Orexin-1 receptor signaling increases motivation for cocaine-associated cues. *The European Journal of Neuroscience*, *41*(9), 1149–1156. https://doi.org/10.1111/ejn.12866.

Bernstein, D. L., Badve, P. S., Barson, J. R., Bass, C. E., & España, R. A. (2017). Hypocretin receptor 1 knockdown in the ventral tegmental area attenuates mesolimbic dopamine signaling and reduces motivation for cocaine. *Addiction Biology*. https://doi.org/10.1111/adb.12553.

Berridge, K. C. (2012). From prediction error to incentive salience: Mesolimbic computation of reward motivation. *The European Journal of Neuroscience*, *35*(7), 1124–1143. https://doi.org/10.1111/j.1460-9568.2012.07990.x.

Borgland, S. L., Chang, S. J., Bowers, M. S., Thompson, J. L., Vittoz, N., Floresco, S. B., et al. (2009). Orexin A/hypocretin-1 selectively promotes motivation for positive reinforcers. *The Journal of Neuroscience*, *29*(36), 11215–11225. https://doi.org/10.1523/JNEUROSCI.6096-08.2009.

Borgland, S. L., Taha, S. A., Sarti, F., Fields, H. L., & Bonci, A. (2006). Orexin A in the VTA is critical for the induction of synaptic plasticity and behavioral sensitization to cocaine. *Neuron*, *49*(4), 589–601. https://doi.org/10.1016/j.neuron.2006.01.016.

Boutrel, B., Kenny, P. J., Specio, S. E., Martin-Fardon, R., Markou, A., Koob, G. F., et al. (2005). Role for hypocretin in mediating stress-induced reinstatement of cocaine-seeking behavior. *Proceedings of the National Academy of Sciences of the United States of America*, *102*(52), 19168–19173. https://doi.org/10.1073/pnas.0507480102.

Brodnik, Z. D., Bernstein, D. L., Prince, C. D., & España, R. A. (2015). Hypocretin receptor 1 blockade preferentially reduces high effort responding for cocaine without promoting sleep. *Behavioural Brain Research*, *291*, 377–384. https://doi.org/10.1016/j.bbr.2015.05.051.

Brown, R. M., Khoo, S. Y., & Lawrence, A. J. (2013). Central orexin (hypocretin) 2 receptor antagonism reduces ethanol self-administration, but not cue-conditioned ethanol-seeking, in ethanol-preferring rats. *The International Journal of Neuropsychopharmacology*, *16*(9), 2067–2079. https://doi.org/10.1017/S1461145713000333.

Brown, R. M., Kim, A. K., Khoo, S. Y., Kim, J. H., Jupp, B., & Lawrence, A. J. (2016). Orexin-1 receptor signalling in the prelimbic cortex and ventral tegmental area regulates cue-induced reinstatement of ethanol-seeking in iP rats. *Addiction Biology*, *21*(3), 603–612. https://doi.org/10.1111/adb.12251.

Burton, A. C., Nakamura, K., & Roesch, M. R. (2015). From ventral-medial to dorsal-lateral striatum: neural correlates of reward-guided decision-making. *Neurobiology of Learning and Memory*, *117*, 51–59. https://doi.org/10.1016/j.nlm.2014.05.003.

Calipari, E. S., & España, R. A. (2012). Hypocretin/orexin regulation of dopamine signaling: implications for reward and reinforcement mechanisms. *Frontiers in Behavioral Neuroscience*, *6*, 54. https://doi.org/10.3389/fnbeh.2012.00054.

Carboni, E., Imperato, A., Perezzani, L., & Di Chiara, G. (1989). Amphetamine, cocaine, phencyclidine and nomifensine increase extracellular dopamine concentrations preferentially in the nucleus accumbens of freely moving rats. *Neuroscience*, *28*(3), 653–661.

Cason, A. M., & Aston-Jones, G. (2013). Role of orexin/hypocretin in conditioned sucrose-seeking in rats. *Psychopharmacology*, *226*(1), 155–165. https://doi.org/10.1007/s00213-012-2902-y.

Cason, A. M., & Aston-Jones, G. (2014). Role of orexin/hypocretin in conditioned sucrose-seeking in female rats. *Neuropharmacology*, *86*, 97–102. https://doi.org/10.1016/j.neuropharm.2014.07.007.

Covey, D. P., Roitman, M. F., & Garris, P. A. (2014). Illicit dopamine transients: Reconciling actions of abused drugs. *Trends in Neurosciences*, *37*(4), 200–210. https://doi.org/10.1016/j.tins.2014.02.002.

de Lecea, L., Kilduff, T. S., Peyron, C., Gao, X., Foye, P. E., Danielson, P. E., et al. (1998). The hypocretins: Hypothalamus-specific peptides with neuroexcitatory activity. *Proceedings of the National Academy of Sciences of the United States of America, 95*(1), 322–327.

Dhaher, R., Hauser, S. R., Getachew, B., Bell, R. L., McBride, W. J., McKinzie, D. L., et al. (2010). The orexin-1 receptor antagonist SB-334867 reduces alcohol relapse drinking, but not alcohol-seeking, in alcohol-preferring (P) rats. *Journal of Addiction Medicine, 4*(3), 153–159. https://doi.org/10.1097/ADM.0b013e3181bd893f.

Di Chiara, G., & Imperato, A. (1988). Drugs abused by humans preferentially increase synaptic dopamine concentrations in the mesolimbic system of freely moving rats. *Proceedings of the National Academy of Sciences of the United States of America, 85*(14), 5274–5278.

España, R. A., Melchior, J. R., Roberts, D. C., & Jones, S. R. (2011). Hypocretin 1/orexin A in the ventral tegmental area enhances dopamine responses to cocaine and promotes cocaine self-administration. *Psychopharmacology, 214*(2), 415–426. https://doi.org/10.1007/s00213-010-2048-8.

España, R. A., Oleson, E. B., Locke, J. L., Brookshire, B. R., Roberts, D. C., & Jones, S. R. (2010). The hypocretin-orexin system regulates cocaine self-administration via actions on the mesolimbic dopamine system. *The European Journal of Neuroscience, 31*(2), 336–348. https://doi.org/10.1111/j.1460-9568.2009.07065.x.

Fadel, J., & Deutch, A. Y. (2002). Anatomical substrates of orexin-dopamine interactions: lateral hypothalamic projections to the ventral tegmental area. *Neuroscience, 111*(2), 379–387.

Farahimanesh, S., Zarrabian, S., & Haghparast, A. (2017). Role of orexin receptors in the ventral tegmental area on acquisition and expression of morphine-induced conditioned place preference in the rats. *Neuropeptides, 66*, 45–51. https://doi.org/10.1016/j.npep.2017.08.003.

Fuchs, R. A., Evans, K. A., Mehta, R. H., Case, J. M., & See, R. E. (2005). Influence of sex and estrous cyclicity on conditioned cue-induced reinstatement of cocaine-seeking behavior in rats. *Psychopharmacology, 179*(3), 662–672. https://doi.org/10.1007/s00213-004-2080-7.

Greenfield, S. F., Back, S. E., Lawson, K., & Brady, K. T. (2010). Substance abuse in women. *The Psychiatric Clinics of North America, 33*(2), 339–355. https://doi.org/10.1016/j.psc.2010.01.004.

Harris, G. C., Wimmer, M., & Aston-Jones, G. (2005). A role for lateral hypothalamic orexin neurons in reward seeking. *Nature, 437*(7058), 556–559. https://doi.org/10.1038/nature04071.

Hollander, J. A., Lu, Q., Cameron, M. D., Kamenecka, T. M., & Kenny, P. J. (2008). Insular hypocretin transmission regulates nicotine reward. *Proceedings of the National Academy of Sciences of the United States of America, 105*(49), 19480–19485. https://doi.org/10.1073/pnas.0808023105.

Hutcheson, D. M., Quarta, D., Halbout, B., Rigal, A., Valerio, E., & Heidbreder, C. (2011). Orexin-1 receptor antagonist SB-334867 reduces the acquisition and expression of cocaine-conditioned reinforcement and the expression of amphetamine-conditioned reward. *Behavioural Pharmacology, 22*(2), 173–181. https://doi.org/10.1097/FBP.0b013e328343d761.

Ikemoto, S. (2007). Dopamine reward circuitry: Two projection systems from the ventral midbrain to the nucleus accumbens-olfactory tubercle complex. *Brain Research Reviews, 56*(1), 27–78. https://doi.org/10.1016/j.brainresrev.2007.05.004.

James, M. H., Charnley, J. L., Levi, E. M., Jones, E., Yeoh, J. W., Smith, D. W., et al. (2011). Orexin-1 receptor signalling within the ventral tegmental area, but not the paraventricular thalamus, is critical to regulating cue-induced reinstatement of cocaine-seeking. *The International Journal of Neuropsychopharmacology, 14*(5), 684–690. https://doi.org/10.1017/s1461145711000423.

Jupp, B., Krivdic, B., Krstew, E., & Lawrence, A. J. (2011). The orexin(1) receptor antagonist SB-334867 dissociates the motivational properties of alcohol and sucrose in rats. *Brain Research, 1391*, 54–59. https://doi.org/10.1016/j.brainres.2011.03.045.

Jupp, B., Krstew, E., Dezsi, G., & Lawrence, A. J. (2011). Discrete cue-conditioned alcohol-seeking after protracted abstinence: Pattern of neural activation and involvement of orexin(1) receptors. *British Journal of Pharmacology, 162*(4), 880–889. https://doi.org/10.1111/j.1476-5381.2010.01088.x.

Kelley, A. E., & Berridge, K. C. (2002). The neuroscience of natural rewards: Relevance to addictive drugs. *The Journal of Neuroscience, 22*(9), 3306–3311.

Khoo, S. Y., & Brown, R. M. (2014). Orexin/hypocretin based pharmacotherapies for the treatment of addiction: DORA or SORA? *CNS Drugs, 28*(8), 713–730. https://doi.org/10.1007/s40263-014-0179-x.

Korotkova, T. M., Sergeeva, O. A., Eriksson, K. S., Haas, H. L., & Brown, R. E. (2003). Excitation of ventral tegmental area dopaminergic and nondopaminergic neurons by orexins/hypocretins. *The Journal of Neuroscience, 23*(1), 7–11.

LeSage, M. G., Perry, J. L., Kotz, C. M., Shelley, D., & Corrigall, W. A. (2010). Nicotine self-administration in the rat: Effects of hypocretin antagonists and changes in hypocretin mRNA. *Psychopharmacology, 209*(2), 203–212. https://doi.org/10.1007/s00213-010-1792-0.

Levy, K. A., Brodnik, Z. D., Shaw, J. K., Perrey, D. A., Zhang, Y., & España, R. A. (2017). Hypocretin receptor 1 blockade produces bimodal modulation of cocaine-associated mesolimbic dopamine signaling. *Psychopharmacology, 234*(18), 2761–2776. https://doi.org/10.1007/s00213-017-4673-y.

Lynch, W. J., & Carroll, M. E. (2000). Reinstatement of cocaine self-administration in rats: Sex differences. *Psychopharmacology, 148*(2), 196–200.

Mahler, S. V., Smith, R. J., & Aston-Jones, G. (2013). Interactions between VTA orexin and glutamate in cue-induced reinstatement of cocaine seeking in rats. *Psychopharmacology, 226*(4), 687–698. https://doi.org/10.1007/s00213-012-2681-5.

Marcus, J. N., Aschkenasi, C. J., Lee, C. E., Chemelli, R. M., Saper, C. B., Yanagisawa, M., et al. (2001). Differential expression of orexin receptors 1 and 2 in the rat brain. *The Journal of Comparative Neurology, 435*(1), 6–25.

Margolis, E. B., Toy, B., Himmels, P., Morales, M., & Fields, H. L. (2012). Identification of rat ventral tegmental area GABAergic neurons. *PLoS One*, 7(7), e42365. https://doi.org/10.1371/journal.pone.0042365.

Nair, S. G., Golden, S. A., & Shaham, Y. (2008). Differential effects of the hypocretin 1 receptor antagonist SB 334867 on high-fat food self-administration and reinstatement of food seeking in rats. *British Journal of Pharmacology, 154*(2), 406–416. https://doi.org/10.1038/bjp.2008.3.

Narita, M., Nagumo, Y., Hashimoto, S., Narita, M., Khotib, J., Miyatake, M., et al. (2006). Direct involvement of orexinergic systems in the activation of the mesolimbic dopamine pathway and related behaviors induced by morphine. *The Journal of Neuroscience, 26*(2), 398–405. https://doi.org/10.1523/JNEUROSCI.2761-05.2006.

Narita, M., Nagumo, Y., Miyatake, M., Ikegami, D., Kurahashi, K., & Suzuki, T. (2007). Implication of protein kinase C in the orexin-induced elevation of extracellular dopamine levels and its rewarding effect. *The European Journal of Neuroscience, 25*(5), 1537–1545. https://doi.org/10.1111/j.1460-9568.2007.05403.x.

Oleson, E. B., & Roberts, D. C. (2012). Cocaine self-administration in rats: Threshold procedures. *Methods in Molecular Biology, 829*, 303–319. https://doi.org/10.1007/978-1-61779-458-2_20.

Plaza-Zabala, A., Flores, A., Martin-Garcia, E., Saravia, R., Maldonado, R., & Berrendero, F. (2013). A role for hypocretin/orexin receptor-1 in cue-induced reinstatement of nicotine-seeking behavior. *Neuropsychopharmacology, 38*(9), 1724–1736. https://doi.org/10.1038/npp.2013.72.

Plaza-Zabala, A., Martin-Garcia, E., de Lecea, L., Maldonado, R., & Berrendero, F. (2010). Hypocretins regulate the anxiogenic-like effects of nicotine and induce reinstatement of nicotine-seeking behavior. *The Journal of Neuroscience*, *30*(6), 2300–2310. https://doi.org/10.1523/JNEUROSCI.5724-09.2010.

Porter-Stransky, K. A., Bentzley, B. S., & Aston-Jones, G. (2017). Individual differences in orexin-I receptor modulation of motivation for the opioid remifentanil. *Addiction Biology*, *22*(2), 303–317. https://doi.org/10.1111/adb.12323.

Prince, C. D., Rau, A. R., Yorgason, J. T., & España, R. A. (2015). Hypocretin/orexin regulation of dopamine signaling and cocaine self-administration is mediated predominantly by hypocretin receptor 1. *ACS Chemical Neuroscience*, *6*(1), 138–146. https://doi.org/10.1021/cn500246j.

Rassnick, S., Pulvirenti, L., & Koob, G. F. (1992). Oral ethanol self-administration in rats is reduced by the administration of dopamine and glutamate receptor antagonists into the nucleus accumbens. *Psychopharmacology*, *109*(1–2), 92–98.

Richards, J. K., Simms, J. A., Steensland, P., Taha, S. A., Borgland, S. L., Bonci, A., et al. (2008). Inhibition of orexin-1/hypocretin-1 receptors inhibits yohimbine-induced reinstatement of ethanol and sucrose seeking in Long-Evans rats. *Psychopharmacology*, *199*(1), 109–117. https://doi.org/10.1007/s00213-008-1136-5.

Roberts, D. C., Bennett, S. A., & Vickers, G. J. (1989). The estrous cycle affects cocaine self-administration on a progressive ratio schedule in rats. *Psychopharmacology*, *98*(3), 408–411.

Roberts, D. C., Brebner, K., Vincler, M., & Lynch, W. J. (2002). Patterns of cocaine self-administration in rats produced by various access conditions under a discrete trials procedure. *Drug and Alcohol Dependence*, *67*(3), 291–299.

Roberts, D. C., Corcoran, M. E., & Fibiger, H. C. (1977). On the role of ascending catecholaminergic systems in intravenous self-administration of cocaine. *Pharmacology, Biochemistry, and Behavior*, *6*(6), 615–620.

Sakurai, T., Amemiya, A., Ishii, M., Matsuzaki, I., Chemelli, R. M., Tanaka, H., et al. (1998). Orexins and orexin receptors: A family of hypothalamic neuropeptides and G protein-coupled receptors that regulate feeding behavior. *Cell*, *92*(4), 573–585.

Salamone, J. D., Correa, M., Mingote, S. M., & Weber, S. M. (2005). Beyond the reward hypothesis: Alternative functions of nucleus accumbens dopamine. *Current Opinion in Pharmacology*, *5*(1), 34–41. https://doi.org/10.1016/j.coph.2004.09.004.

SAMHSA (2014). *Results from the 2013 national survey on drug use and health: Summary of national findings*. Rockville, MD: Substance Abuse and Mental Health Services Administration.

Sartor, G. C., & Aston-Jones, G. S. (2012). A septal-hypothalamic pathway drives orexin neurons, which is necessary for conditioned cocaine preference. *Journal of Neuroscience*, *32*(13), 4623–4631.

Schmeichel, B. E., Barbier, E., Misra, K. K., Contet, C., Schlosburg, J. E., Grigoriadis, D., et al. (2015). Hypocretin receptor 2 antagonism dose-dependently reduces escalated heroin self-administration in rats. *Neuropsychopharmacology*, *40*(5), 1123–1129. https://doi.org/10.1038/npp.2014.293.

Shaw, J. K., Ferris, M. J., Locke, J. L., Brodnik, Z. D., Jones, S. R., & España, R. A. (2016). Hypocretin/orexin knock-out mice display disrupted behavioral and dopamine responses to cocaine. *Addiction Biology*, *22*(6), 1695–1705. https://doi.org/10.1111/adb.12432.

Smith, R. J., & Aston-Jones, G. (2012). Orexin/hypocretin 1 receptor antagonist reduces heroin self-administration and cue-induced heroin seeking. *The European Journal of Neuroscience*, *35*(5), 798–804. https://doi.org/10.1111/j.1460-9568.2012.08013.x.

Smith, R. J., Tahsili-Fahadan, P., & Aston-Jones, G. (2010). Orexin/hypocretin is necessary for context-driven cocaine-seeking. *Neuropharmacology*, *58*(1), 179–184. https://doi.org/10.1016/j.neuropharm.2009.06.042.

Spanagel, R., & Weiss, F. (1999). The dopamine hypothesis of reward: Past and current status. *Trends in Neurosciences*, *22*(11), 521–527.

Tung, L. W., Lu, G. L., Lee, Y. H., Yu, L., Lee, H. J., Leishman, E., et al. (2016). Orexins contribute to restraint stress-induced cocaine relapse by endocannabinoid-mediated disinhibition of dopaminergic neurons. *Nature Communications*, 7, 12199. https://doi.org/10.1038/ncomms12199.

Tzschentke, T. M. (1998). Measuring reward with the conditioned place preference paradigm: A comprehensive review of drug effects, recent progress and new issues. *Progress in Neurobiology*, *56*(6), 613–672.

Tzschentke, T. M. (2007). Measuring reward with the conditioned place preference (CPP) paradigm: Update of the last decade. *Addiction Biology*, *12*(3-4), 227–462. https://doi.org/10.1111/j.1369-1600.2007.00070.x.

Vittoz, N. M., & Berridge, C. W. (2006). Hypocretin/orexin selectively increases dopamine efflux within the prefrontal cortex: Involvement of the ventral tegmental area. *Neuropsychopharmacology*, *31*(2), 384–395. https://doi.org/10.1038/sj.npp.1300807.

Voorhees, C. M., & Cunningham, C. L. (2011). Involvement of the orexin/hypocretin system in ethanol conditioned place preference. *Psychopharmacology*, *214*(4), 805–818. https://doi.org/10.1007/s00213-010-2082-6.

Wang, B., You, Z. B., & Wise, R. A. (2009). Reinstatement of cocaine seeking by hypocretin (orexin) in the ventral tegmental area: independence from the local corticotropin-releasing factor network. *Biological Psychiatry*, *65*(10), 857–862. https://doi.org/10.1016/j.biopsych.2009.01.018.

Zhou, L., Ghee, S. M., Chan, C., Lin, L., Cameron, M. D., Kenny, P. J., et al. (2012). Orexin-1 receptor mediation of cocaine seeking in male and female rats. *The Journal of Pharmacology and Experimental Therapeutics*, *340*(3), 801–809. https://doi.org/10.1124/jpet.111.187567.

CHAPTER 6

Orexin and Cocaine Addiction

Alessandra Matzeu, Rémi Martin-Fardon
Department of Neuroscience, The Scripps Research Institute, La Jolla, CA, United States

OVERVIEW

Cocaine addiction, similar to other drug addictions, is a chronic relapsing disorder that is characterized by compulsive drug seeking that persists despite adverse consequences (American Psychiatric Association, and American Psychiatric Association, DSM-5 Task Force, 2013). No pharmacotherapy is currently approved for the treatment of cocaine use disorder, despite the substantial effort that has been expended by the scientific community to investigate and develop effective compounds. Addiction requires significant medical attention and requires the development of novel pharmacotherapies. Animal models (e.g., self-administration and reinstatement) have provided insights into the neurobiological basis of cocaine addiction and particularly relapse (i.e., drug-seeking behavior in preclinical research settings). Alterations of neurochemical function, signal transduction, and transcription mechanisms in sensitive neuronal structures may mediate the chronic actions of drug that lead to vulnerability to relapse and may provide essential insights into the neuroadaptations that are associated with drug addiction.

Studies have explored the neurocircuitry that mediates the reinstatement of cocaine-seeking behavior, identifying a core circuit with interconnected cortical and limbic regions that participates in the regulation of cocaine seeking. Major components of this circuitry include the ventral tegmental area (VTA), central nucleus of the amygdala (CeA), bed nucleus of the stria terminalis (BNST), nucleus accumbens (NAc), basolateral amygdala, hippocampus, medial prefrontal cortex (mPFC), and dorsal striatum (Everitt, Dickinson, & Robbins, 2001; Ito, Dalley, Robbins, & Everitt, 2002; Kalivas & Volkow, 2005; McFarland & Kalivas, 2001; See, 2002). Neural circuits that are implicated in drug conditioning, craving, and relapse overlap with those that are involved in natural reward seeking. Neuroadaptations of

The Orexin/Hypocretin System
https://doi.org/10.1016/B978-0-12-813751-2.00006-1

these neuronal systems are believed to be responsible for the maladaptive (compulsive) nature of addiction (Aston-Jones & Harris, 2004; Kalivas & O'Brien, 2008; Kelley & Berridge, 2002; Wanat, Willuhn, Clark, & Phillips, 2009).

Independently discovered in 1998 by Sakurai (Sakurai et al., 1998) and de Lecea (de Lecea et al., 1998), orexin A (OrxA) and orexin B (OrxB), also referred to as hypocretin 1 (Hcrt1) and hypocretin 2 (Hcrt2), are neuropeptides that derive from a common precursor, prepro-orexin (ppOrx), that is produced exclusively in well-defined subregions of the hypothalamus (Fig. 1A): lateral hypothalamus (LH), dorsomedial hypothalamus (DMH), and perifornical area (PFA; de Lecea et al., 1998; Peyron et al., 1998; Sakurai et al., 1998). Orexins bind two G-protein-coupled receptors (Gotter, Webber, Coleman, Renger, & Winrow, 2012): orexin receptor 1 (OrxR1) and orexin receptor 2 (OrxR2). OrxR1 binds OrxA with 20–30 nM affinity but has much lower affinity (10- to 1000-fold lower) for OrxB, whereas OrxR2 binds both peptides with similar affinity (in the 40-nM range; (Sakurai et al., 1998; Scammell & Winrow, 2011). The two Orx receptors are distributed throughout the brain, but some overlap between OrxR1 and OrxR2 mRNA expression has been reported (Marcus et al., 2001; Trivedi, Yu, MacNeil, Van der Ploeg, & Guan, 1998). For example, the prefrontal cortex predominantly expresses OrxR1 mRNA, whereas the NAc mainly expresses OrxR2 mRNA (e.g., Marcus et al., 2001), suggesting different physiological roles for each receptor subtype. In fact, a functional difference between OrxR1 and OrxR2 has been suggested (Aston-Jones et al., 2010), with OrxR1 signaling involved in reward seeking and OrxR2 signaling involved in arousal.

In situ hybridization studies have shown that ppOrx mRNA is expressed in a restricted area of the brain (de Lecea et al., 1998; Gautvik et al., 1996), but Orx projections are widely distributed throughout the brain, with particularly dense projections to the tuberomammillary nucleus (TMN), paraventricular nucleus of the thalamus (PVT), arcuate nucleus of the hypothalamus (ARC), locus coeruleus (LC; which receives the densest Orx inputs), and raphe nucleus (RN; Fig. 1B; Date et al., 1999; Nambu et al., 1999; Peyron et al., 1998; Sakurai, 2007; Sakurai & Mieda, 2011). Additionally, Orx neurons project to the mPFC and mesolimbic system, such as the VTA and NAc shell (NAcsh; Li, Hu, & de Lecea, 2014). Studies that used anterograde and retrograde tracers suggested that Orx neurons are innervated by the mPFC, NAcsh, PVT, amygdala, BNST, ARC, and preoptic area (POA; Fig. 1B; Kirouac, 2015; Sakurai et al., 2005).

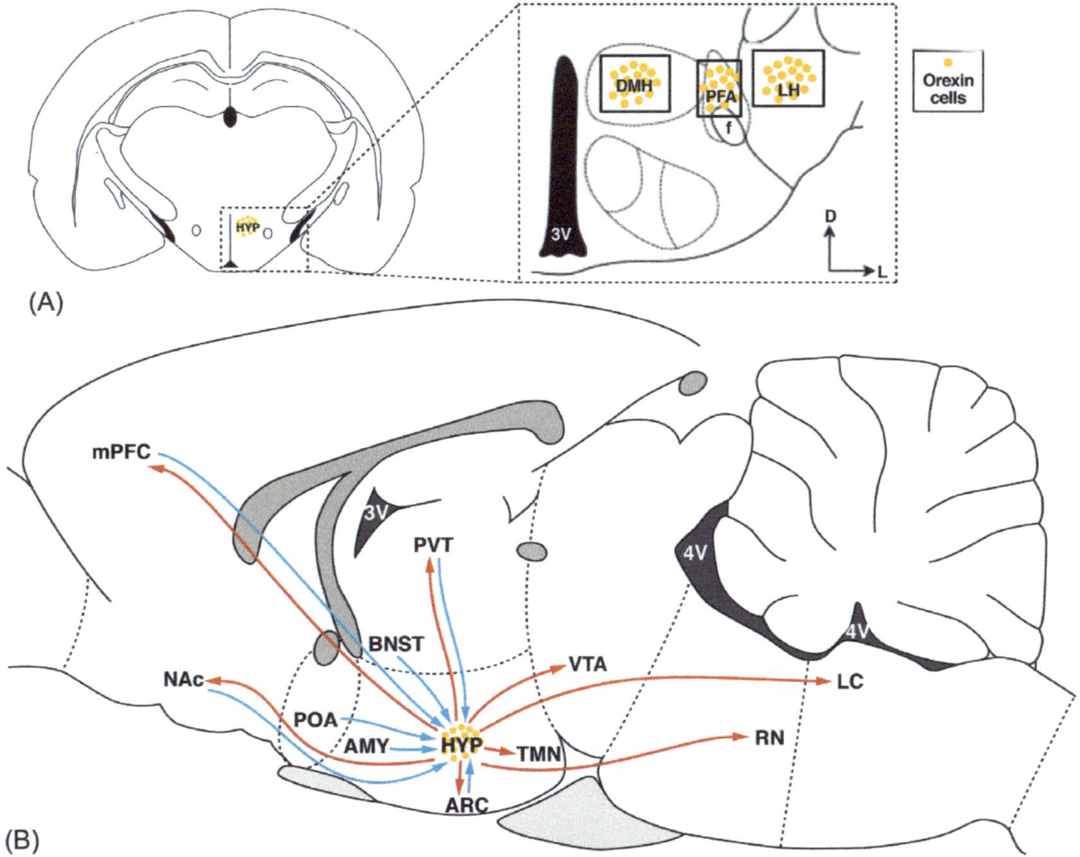

Fig. 1 (A) Schematic illustration of the rostrocaudal level of orexin neurons in the hypothalamus. *LH*, lateral hypothalamus; *DMH*, dorsomedial hypothalamus; *PFA*, perifornical area (de Lecea et al., 1998; Sakurai et al., 1998). (B) Schematic representation of major hypothalamic orexin neurons (HYP) efferents *(red)* (Date et al., 1999; Li et al., 2014; Nambu et al., 1999; Peyron et al., 1998; Sakurai & Mieda, 2011) and afferents *(blue)* (Kirouac, 2015; Sakurai et al., 2005) within the neurocircuitry of addiction. *HYP*, hypothalamus; *TMN*, tuberomammillary nucleus; *PVT*, paraventricular nucleus of the thalamus; *ARC*, arcuate nucleus of the hypothalamus; *LC*, locus coeruleus; *RN*, raphe nucleus; *mPFC*, medial prefrontal cortex; *VTA*, ventral tegmental area; *AMY*, amygdala; *POA*, preoptic area; *BNST*, bed nucleus of the stria terminalis; *NAc*, nucleus accumbens.

As a result of its projections throughout the brain, the Orx system is involved in numerous physiological functions, such as the regulation of feeding, arousal, sleep/wake states, stress responses, energy homeostasis, and reward (for review, see Tsujino & Sakurai, 2013). The Orx system is known to be recruited by both conventional reinforcers (Cason & Aston-Jones, 2013a, 2013b; Olney, Navarro, & Thiele, 2015) and drugs of abuse (Harris, Wimmer, & Aston-Jones, 2005; Mahler, Smith, Moorman, Sartor, & Aston-Jones, 2012), including cocaine (Boutrel et al., 2005; España et al., 2010; Hollander, Pham, Fowler, & Kenny, 2012; Martin-Fardon and Weiss, 2014; Muschamp et al., 2014; Prince, Rau, Yorgason, & España, 2015; Schmeichel, Herman, Roberto, & Koob, 2017; Smith, See, & Aston-Jones, 2009; Smith, Tahsili-Fahadan, & Aston-Jones, 2010; Wang, You, & Wise, 2009; Zhou, Smith, Do, Aston-Jones, & See, 2012).

The present chapter discusses significant preclinical studies that demonstrate the participation of the Orx system in cocaine addiction and examines the ways in which pharmacological manipulations of the Orx system may potentially reduce drug intake and prevent craving and relapse.

PHARMACOLOGICAL TOOLS TO TARGET OrxRs: SORAs AND DORAs

Several single OrxR antagonists (SORAs) have been synthesized that target OxR1. These compounds include *N*-(2-methyl-6-benzoxazolyl)-*N*'-1,5-naphthyridin-4-yl urea (SB334867; Porter et al., 2001; Smart et al., 2001), [5-(2-fluorophenyl)-2-methyl-1,3-thiazol-4-yl]-[(2*S*)-2-[(5-phenyl-1,3,4-oxadiazol-2-yl)methyl]pyrrolidin-1-yl]methanone (SB674042; (Langmead et al., 2004), 1-(6,8-difluoro-2-methylquinolin-4-yl)-3-[4-(dimethylamino)phenyl]urea (SB408124; (Langmead et al., 2004), 1-(5,8-difluoroquinolin-4-yl)-3-(4-dimethylamino-phenyl)-urea (SB410220; (Langmead et al., 2004; Porter et al., 2001), [(2*S*,5*S*)-2-[([5-bromo-2-pyridyl]amino) methyl]-5-methyl-1-piperidyl]-(3-fluoro-2-methoxy-phenyl)methanone (GSK1059865; (Gozzi et al., 2011), and (a*R*,1*S*)-1-[(3,4-dimethoxyphenyl) methyl]-3,4-dihydro-6,7-dimethoxy-*N*-(1-methylethyl)-a-phenyl-2(1*H*)-isoquinolineacetamide (ACT335827; Steiner, Gatfield, et al., 2013). Most of these SORAs have at least 50-fold selectivity for OrxR1 over OrxR2 (Scammell & Winrow, 2011; Zhou, Sun, & See, 2011). Among these, SB334867 has been the most researched compound because of its high selectivity for OrxR1, potency, and bioavailability (Scammell & Winrow, 2011; Zhou et al., 2011). SORAs that target OrxR2 have also been developed. Some of these compounds, such as *N*-(2,4-dibromophenyl)-*N*'-[(4*S*,5*S*)-2,2-dimethyl-4-phenyl-1,3-dioxan-5-yl]-urea (JNJ10397049; (Dugovic et al., 2009; McAtee et al., 2004), *N*-ethyl-2-[(6-methoxypyridin-3-yl)-(2-methylphenyl)sulfonylamino]-*N*-(pyridin-3-ylmethyl)acetamide (EMPA; Malherbe et al., 2009), [4-pyridylmethyl-(*S*)-*tert*-leucyl-6,7-dimethoxy-1,2,3,4-tetrahydroisoquinoline (4PT; (Smith et al., 2009), and (2*S*)-1-(3,4-dihydro-6,7-dimethoxy-2(1*H*)-isoquinolinyl)-3,3-dimethyl-2-[(4-pyridinylmethyl)amino]-1-butanone hydrochloride (TCSOX229; Hirose et al., 2003), have at least 250-fold selectivity for OrxR2 over OrxR1 (Scammell & Winrow, 2011; Zhou et al., 2011) and have been studied because of their ability to decrease several addiction-related behaviors. However, OrxR2 SORAs have less consistent effects on drug-motivated behaviors compared with OrxR1 SORAs, especially SB334867. In addition to SORAs, several dual OrxR antagonists (DORAs) have been developed for the treatment of insomnia. These DORA compounds include, but are not limited to, almorexant ((2*R*)-2-[(1*S*)-6,7-dimethoxy-1-[2-[4-(trifluoromethyl)phenyl] ethyl]-3,4-dihydro-1*H*-isoquinolin-2-yl]-*N*-methyl-2-phenylacetamide (ACT078573; Brisbare-Roch et al., 2007; Malherbe, Borroni, Pinard, Wettstein, & Knoflach, 2009), suvorexant ([(7*R*)-4-(5-chloro-1,3-benzoxazol-2-yl)-7-methyl-1,4-diazepan-1-yl]-[5-methyl-2-(triazol-2-yl)phenyl]methanone (MK4305; Cox & Garbaccio, 2010), Merck DORA-1 (Bergman et al., 2008), Merck DORA-5 (Whitman et al., 2009), and *N*-([(2*S*)-1-([5-(4-fluorophenyl)-2-methyl-4-thiazolyl] carbonyl)-2-piperidinyl]methyl)-4-benzofurancarboxamide (SB649868; Renzulli et al., 2011). To date, the only studies that have examined the effects of DORAs on addiction-related behaviors have used exclusively almorexant (Hoch et al., 2013; LeSage, Perry, Kotz, Shelley, & Corrigall, 2010; Steiner, Lecourt, and Jenck, 2013). Notably, the majority of preclinical studies that investigated the effects of OrxR blockade used acute administration of OrxR antagonists, although repeated inhibition of OrxR1-mediated signaling under certain conditions exerts a lasting and specific prevention of cocaine-seeking behavior that is induced by drug-related cues (Zhou et al., 2012). The Orx system is involved in numerous physiological processes, and Orx neurons interface with multiple neural systems. The repeated inhibition of OrxRs likely can affect several neural pathways and thus produce significant nonspecific side effects. For example, Orx dysfunction produces narcolepsy (Chemelli et al., 1999; Lin et al., 1999; Peyron et al., 2000; Thannickal et al., 2000), and pharmacological manipulation of the Orx system can affect food intake (e.g., (Haynes et al., 2000; Sakurai et al., 1998). These are important limitations that will need to be carefully evaluated when considering clinical treatment with specific OrxR antagonists.

MANIPULATION OF THE OREXIN SYSTEM AND COCAINE ADDICTION: BEHAVIORAL DATA

Studies that investigated cocaine behavioral sensitization and conditioned place preference (CPP) have shown that the peripheral administration of SB334867 or local injection in the VTA blocked the acquisition but not expression of cocaine sensitization (Borgland, Taha, Sarti, Fields, & Bonci, 2006) and attenuated the acquisition and expression of cocaine-induced CPP (Hutcheson et al., 2011), suggesting that OrxR1 blockade protects against the acquisition of cocaine sensitization and cocaine-induced CPP. Furthermore, targeting both OrxR1 and OrxR2 with almorexant was shown to attenuate the expression of cocaine-induced CPP at high doses, with no effect on the expression of behavioral sensitization (Steiner, Lecourt, et al., 2013).

The use of voluntary drug self-administration has allowed characterization of the effect of targeting the Orx system using more relevant models of addiction. Several studies have implicated the Orx system in cocaine self-administration, depending on the schedule of reinforcement that is used. For example, central (i.e., intra-VTA or intracerebroventricular [ICV]) administration of OrxA and systemic or intra-VTA administration of SB334867 had no effect on cocaine intake under a fixed-ratio 1 (FR-1) schedule of reinforcement (Boutrel et al., 2005; Espana et al., 2010; Hutcheson et al., 2011; Smith et al., 2009). Noticeable effects are observed, however, when high effort (i.e., work) is required for the animals to obtain the drug, such as under an FR-5 schedule, during which systemic SB334837 administration decreased cocaine self-administration (Hollander et al., 2012). Other studies showed that systemic or intra-VTA administration of SB334837 reduced the breakpoint for cocaine self-administration on a progressive-ratio (PR) schedule of reinforcement and decreased cocaine intake in a discrete trial (DT) procedure (Borgland et al., 2009; Espana et al., 2010). Intra-VTA OrxA administration increased the motivation to obtain cocaine in a PR schedule and cocaine self-administration on a DT schedule (Espana, Melchior, Roberts, & Jones, 2011). SB334867 dose dependently reduced the stimulatory effects of cocaine on brain reward systems (i.e., lowered intracranial self-stimulation [ICSS] thresholds), and OrxR1 knockout mice self-administered less cocaine than wild-type mice (Hollander et al., 2012). Moreover, in cocaine self-administering rats, an intra-VTA injection of SB334867 attenuated basal and cocaine-induced extracellular dopamine levels in the NAc, whereas an intra-VTA injection of OrxA increased both basal and cocaine-induced extracellular dopamine levels in the NAc (Espana et al., 2010, 2011). These findings suggest that Orx transmission has an essential function in regulating the reinforcing actions of cocaine and modulating the effects of cocaine on the mesolimbic dopamine system.

Using a viral vector approach to knock down the hypothalamic production of Orx, our group showed that chronic *Orx* silencing did not modify cocaine self-administration when rats self-administered cocaine under short-access conditions (2 h/day) but robustly attenuated cocaine intake when rats self-administered cocaine under long-access (LgA) conditions (6 h/day). The latter model mimics key features of compulsive cocaine taking (e.g., (Ahmed & Koob, 1998). Additionally, *Orx* silencing decreased the motivation for both cocaine and palatable food (i.e., sweetened condensed milk [SCM]) under a PR schedule of reinforcement (i.e., high effort) but did not alter responding for SCM under a FR schedule. Importantly, *Orx* silencing did not affect the self-administration of regular food pellets or water and did not have nonspecific effects on general measures of arousal-dependent behaviors (e.g., locomotor activity, anxiety-like behavior, and stress-induced analgesia; (Schmeichel et al., 2018). These original data support the hypothesis that Orx neurotransmission promotes operant responding for both drug and nondrug rewards under conditions that require a high degree of motivation. Moreover, the data show that *Orx* silencing decreased cocaine self-administrating in rats that had extended access, providing compelling evidence of involvement of the Orx system in compulsive-like cocaine taking (Schmeichel et al., 2018).

In addition to its involvement in the reinforcing actions of cocaine, Orx has been directly implicated in cocaine-seeking behavior. The central administration of OrxA (ICV, intra-VTA, or intra-PVT) reinstated cocaine-seeking behavior (Boutrel et al., 2005; Matzeu, Kerr, Weiss, & Martin-Fardon, 2016; Wang et al., 2009). Peripheral or intra-VTA injections of SB334867 and an intra-VTA injection of SB334867 prevented cocaine-seeking behavior that was induced by drug-related stimuli, drug contexts, and footshock stress (Boutrel et al., 2005; James et al., 2011; Mahler, Smith, & Aston-Jones, 2013; Smith et al., 2009, 2010). Interestingly, the blockade of OrxR1 by peripheral SB334867 administration dose dependently and selectively reversed conditioned reinstatement that was induced by cocaine-related stimuli, without affecting reward seeking that was produced by the same stimulus when it was conditioned to a highly palatable food (i.e., SCM; Martin-Fardon & Weiss, 2014). Moreover, the intra-VTA administration of OrxB (which mainly acts on OrxR2) had no effect on cocaine-seeking behavior (Wang et al., 2009), and systemic 4PT administration had no effect on the cue-induced reinstatement of cocaine seeking (Smith et al., 2009), suggesting that OrxR1 blockade may be beneficial for the prevention of cocaine craving and relapse, without affecting behavior that is directed toward natural rewards.

The mesocorticolimbic system, especially in the VTA, has been the focus of most research that implicates Orx in addiction, consequently neglecting other important systems. Recent studies have investigated the participation of other brain regions that receive Orx inputs. For example, the PVT receives Orx projections (Kirouac, Parsons, & Li, 2005; Parsons, Li, & Kirouac, 2006), and a recent study showed that an injection of OrxA in the PVT reinstated (primed) extinguished cocaine (LgA)- and SCM-seeking behavior (Matzeu, Kallupi, George, Schweitzer, & Martin-Fardon, 2017). However, although OrxA reinstated both cocaine and SCM seeking, it produced a stronger reinstatement of cocaine vs. SCM seeking, suggesting the cocaine-induced dysregulation of Orx-PVT transmission following cocaine exposure (Matzeu et al., 2017). Importantly, OrxA's priming effects on cocaine seeking were selectively reversed by TCSOX229 but not by SB334867 (Matzeu, Kerr, et al., 2016). These data suggest that the participation of Orx transmission in drug seeking is not exclusive to the VTA but also depends on other brain regions, and different receptors (e.g., OrxR1, OrxR2, or both) could be involved in mediating drug seeking.

Altogether, these findings indicate a role for OrxA in cocaine-induced locomotor sensitization, conditioned reinstatement, and the primary reinforcing actions of cocaine when a high level of effort and motivation are necessary to obtain the drug. Moreover, the blockade of OrxR1 by a peripheral injection of an OrxR1 antagonist plays a key role in appetitive behavior that is controlled by reward-related stimuli with selectivity for cocaine seeking (vs. SCM), identifying OrxR1 as a potential treatment target for cocaine relapse prevention. The blockade of OrxR2 by discrete injections in the PVT prevented OrxA-induced reinstatement, which may have important implications. This suggests that both receptors could be targeted to prevent craving and relapse, resulting in a more efficient pharmacotherapy that targets several brain regions that are affected by cocaine use disorder.

CELLULAR AND MOLECULAR EVIDENCE OF COCAINE-INDUCED NEUROADAPTATION OF THE OREXIN SYSTEM

Voluntary (self-administered) or involuntary (experimenter-administered) cocaine exposure induces excitatory synapse remodeling, consequently increasing excitatory inputs to Orx neurons in the hypothalamus (Yeoh et al., 2012), a phenomenon that may contribute to the persistence of drug-seeking behavior and relapse. In the VTA, cocaine-induced postsynaptic plasticity can be facilitated in rats by OrxA and prevented by SB334867 (Borgland et al., 2006). Glutamatergic transmission and plasticity in dopaminergic VTA neurons have previously been reported to be important for learning drug–environment associations, such as in the learning (acquisition) phase of cocaine-induced CPP (Harris & Aston-Jones, 2003). In the VTA, OrxA potentiated glutamate signaling postsynaptically to promote synaptic plasticity (Borgland et al., 2006), whereas OrxB potentiated glutamatergic signaling pre- and postsynaptically (Borgland, Storm, & Bonci, 2008), suggesting that Orx in the VTA plays an important role in the synaptic plasticity of dopaminergic neurons that is important for drug-related behaviors.

In conclusion, evidence of the critical contribution of Orx to drug motivation originated from findings that showed a correlation between cocaine-induced CPP and Orx neuron activation in the LH but not DMH or PFA (Harris et al., 2005). Recently, Martin-Fardon, Cauvi, Kerr, and Weiss (2016) showed that Orx neurons in the LH, DMH, and PFA were preferentially activated following the presentation of cocaine-related stimuli compared with SCM-related stimuli, an effect that persisted over time and supported a role for the Orx system in perseverative, compulsive-like cocaine seeking but not behavior that is motivated by palatable food. Moreover, an analysis of Orx recruitment patterns suggested that the failure of Orx neurons in LH to respond to inhibitory inputs from Orx neurons in the DMH/PFA—most likely through GABA and/or glycine neurotransmission (e.g., Porrino, Coons, & MacGregor, 1983)—may contribute to the perseverative nature of cocaine seeking (Martin-Fardon et al., 2016). An injection of OrxA in the PVT produced significant cocaine-seeking behavior that was associated with strong activation (Fos^+) of Orx neurons (Orx^+) in the LH/DMH/PFA (Matzeu, Weiss, & Martin-Fardon, 2016).

Remarkably, a reduction of ppOrx mRNA expression in the LH was observed following cocaine-induced CPP, suggesting a compensatory feedback mechanism that follows robust neuronal activation that is induced by cocaine (Zhou et al., 2008). Overall, these findings strongly support the hypothesis that a history of cocaine use perturbs the "normal" function of the Orx system that can, in turn, be one of the causes of the compulsive nature of cocaine seeking.

GLOSSARY

Conditioned Place Preference A Pavlovian (classical) conditioning procedure that is used to measure the motivational effects of experiences (e.g., drugs). This can also be used to measure conditioned place aversion (CPA) with an identical procedure that instead involves aversive stimuli. Specifically, CPP and CPA are commonly used with rodents to study the affective (positive, neutral, or negative) effects of drugs. During conditioning, a distinct environment is paired with drug injections, and another context is paired with vehicle injections. During a subsequent drug-free test, the animal chooses between the drug- and vehicle-paired environments. An increase in the time spent in the drug-paired environment is a measure of the rewarding effects of a drug, whereas a decrease or avoidance is a measure of aversion.

Conventional Reinforcer or Natural Reward The terms "conventional reinforcer" and "natural reward" are often used interchangeably to define a nondrug condition (usually a sweet, highly palatable solution) that serves as a comparison control for the drug.

Cue-Induced Reinstatement/Conditioned Reinstatement of Drug-Seeking Behavior A paradigm in which rats are trained to self-administer a drug with the presence of cues that are paired with drug availability. Following extinction, during which the cues and drug are absent, presentation of the cues that were associated with drug availability reinstates extinguished drug-seeking behavior.

Extinction In self-administration studies, extinction means the decline of an operant response when the reinforcer is withdrawn.

Positive Reinforcement A procedure in which the instrumental response (e.g., a lever press) is emitted to obtain a reinforcer (e.g., cocaine, palatable food) that will result in pleasurable (rewarding) effects.

Prime-Induced Reinstatement Noncontingent injections of the self-administered drug reinstate extinguished drug-seeking behavior.

Reinstatement In preclinical studies, reinstatement (relapse in humans) refers to the resumption of drug-seeking behavior after extinction. Reinstatement can be induced by a drug prime, drug-related stimuli, or stress.

Relapse The resumption of drug taking following a period of voluntary or involuntary abstinence in humans. Drug priming (i.e., a small amount of the drug), drug-related stimuli, and stress are three major factors that can induce intense craving in abstinent individuals and thus lead to relapse.

Self-Administration The model of intravenous drug self-administration in animals has been used to characterize whether different psychostimulants support self-administration. Drug self-administration in animals may be a predictor of abuse liability in humans. It usually consists of training an animal to voluntary press a lever to obtain an injection of a drug. Fixed-ratio (FR) schedules require a predefined number of operant responses to dispense one unit of reinforcer. Standard FR reinforcement schedules include FR1, FR5, and FR10, which require 1, 5, and 10 operant responses to dispense a unit of reinforcer, respectively. Under a progressive-ratio (PR) schedule of reinforcement, following each infusion of the drug, the work load (ratio) progressively increases. The breakpoint is defined as the final/highest ratio completed during each session. The breakpoint reflects the motivation of an animal to obtain the drug and is thus an indicator of the reinforcing efficacy of a drug. The discrete trials procedure is an extension of an FR schedule that allows 24-h access to a drug but limits the number of injections an animal can receive each hour. For example, during a discrete trials session, rats may have the opportunity to self-administer one dose of cocaine during 10-min trials at 20-min intervals (two trials per hour, 24 h/day). Trials began with the presentation of a cocaine-associated lever and are terminated once the rat depresses the lever or after 10 min. This schedule is well suited for use over 24-h periods and can provide insights into diurnal variability in self-administration behavior that allows for optimal sensitivity for measuring the efficacy of agents in increasing or decreasing drug self-administration.

Sensitization (Locomotor Sensitization) One of the effects of psychomotor stimulants is that repeated administration of these drugs enhances many of their effects. This phenomenon is known as sensitization or reverse tolerance. Early findings with cocaine demonstrated that repetitive intermittent treatment augmented hyperactivity in different species. Today, these effects have been well documented. Generally, these studies have shown that with repeated administration, cocaine increases locomotor activity and the intensity of stereotyped behavior. Sensitization may play a role in addiction or other disorders, such as drug craving and psychostimulant-induced psychosis.

ACKNOWLEDGMENTS

This is publication number 29620 from The Scripps Research Institute. This work was supported by the National Institute on Drug Abuse and National Institute on Alcohol Abuse and Alcoholism (grant no. DA033344, AA024146, AA006420, and AA022249). The authors thank M. Arends for assistance with manuscript preparation.

REFERENCES

Ahmed, S. H., & Koob, G. F. (1998). Transition from moderate to excessive drug intake: change in hedonic set point. *Science*, *282*(5387), 298–300.

American Psychiatric Association, & American Psychiatric Association, DSM-5 Task Force. (2013). *Diagnostic and statistical manual of mental disorders: DSM-5* (5th ed.). Washington, DC: American Psychiatric Association.

Aston-Jones, G., & Harris, G. C. (2004). Brain substrates for increased drug seeking during protracted withdrawal. *Neuropharmacology*, *47*(Suppl 1), 167–179. https://doi.org/10.1016/j.neuropharm.2004.06.020.

Aston-Jones, G., Smith, R. J., Sartor, G. C., Moorman, D. E., Massi, L., Tahsili-Fahadan, P., et al. (2010). Lateral hypothalamic orexin/hypocretin neurons: a role in reward-seeking and addiction. *Brain Research, 1314*, 74–90. https://doi.org/10.1016/j.brainres.2009.09.106.

Bergman, J. M., Roecker, A. J., Mercer, S. P., Bednar, R. A., Reiss, D. R., Ransom, R. W., et al. (2008). Proline bis-amides as potent dual orexin receptor antagonists. *Bioorganic & Medicinal Chemistry Letters, 18*(4), 1425–1430. https://doi.org/10.1016/j.bmcl.2008.01.001.

Borgland, S. L., Chang, S. J., Bowers, M. S., Thompson, J. L., Vittoz, N., Floresco, S. B., et al. (2009). Orexin A/hypocretin-1 selectively promotes motivation for positive reinforcers. *The Journal of Neuroscience, 29*(36), 11215–11225. https://doi.org/10.1523/JNEUROSCI.6096-08.2009.

Borgland, S. L., Storm, E., & Bonci, A. (2008). Orexin B/hypocretin 2 increases glutamatergic transmission to ventral tegmental area neurons. *The European Journal of Neuroscience, 28*(8), 1545–1556. https://doi.org/10.1111/j.1460-9568.2008.06397.x.

Borgland, S. L., Taha, S. A., Sarti, F., Fields, H. L., & Bonci, A. (2006). Orexin A in the VTA is critical for the induction of synaptic plasticity and behavioral sensitization to cocaine. *Neuron, 49*(4), 589–601. https://doi.org/10.1016/j.neuron.2006.01.016.

Boutrel, B., Kenny, P. J., Specio, S. E., Martin-Fardon, R., Markou, A., Koob, G. F., et al. (2005). Role for hypocretin in mediating stress-induced reinstatement of cocaine-seeking behavior. *Proceedings of the National Academy of Sciences of the United States of America, 102*(52), 19168–19173. https://doi.org/10.1073/pnas.0507480102.

Brisbare-Roch, C., Dingemanse, J., Koberstein, R., Hoever, P., Aissaoui, H., Flores, S., et al. (2007). Promotion of sleep by targeting the orexin system in rats, dogs and humans. *Nature Medicine, 13*(2), 150–155. https://doi.org/10.1038/nm1544.

Cason, A. M., & Aston-Jones, G. (2013a). Attenuation of saccharin-seeking in rats by orexin/hypocretin receptor 1 antagonist. *Psychopharmacology (Berl), 228*(3), 499–507. https://doi.org/10.1007/s00213-013-3051-7.

Cason, A. M., & Aston-Jones, G. (2013b). Role of orexin/hypocretin in conditioned sucrose-seeking in rats. *Psychopharmacology (Berl), 226*(1), 155–165. https://doi.org/10.1007/s00213-012-2902-y.

Chemelli, R. M., Willie, J. T., Sinton, C. M., Elmquist, J. K., Scammell, T., Lee, C., et al. (1999). Narcolepsy in orexin knockout mice: molecular genetics of sleep regulation. *Cell, 98*(4), 437–451.

Cox, C. D., & Garbaccio, R. M. (2010). Discovery of allosteric inhibitors of kinesin spindle protein (KSP) for the treatment of taxane-refractory cancer: MK-0731 and analogs. *Anti-Cancer Agents in Medicinal Chemistry, 10*(9), 697–712.

Date, Y., Ueta, Y., Yamashita, H., Yamaguchi, H., Matsukura, S., Kangawa, K., et al. (1999). Orexins, orexigenic hypothalamic peptides, interact with autonomic, neuroendocrine and neuroregulatory systems. *Proceedings of the National Academy of Sciences of the United States of America, 96*(2), 748–753.

de Lecea, L., Kilduff, T. S., Peyron, C., Gao, X., Foye, P. E., Danielson, P. E., et al. (1998). The hypocretins: hypothalamus-specific peptides with neuroexcitatory activity. *Proceedings of the National Academy of Sciences of the United States of America, 95*(1), 322–327.

Dugovic, C., Shelton, J. E., Aluisio, L. E., Fraser, I. C., Jiang, X., Sutton, S. W., et al. (2009). Blockade of orexin-1 receptors attenuates orexin-2 receptor antagonism-induced sleep promotion in the rat. *The Journal of Pharmacology and Experimental Therapeutics, 330*(1), 142–151. https://doi.org/10.1124/jpet.109.152009.

Espana, R. A., Melchior, J. R., Roberts, D. C., & Jones, S. R. (2011). Hypocretin 1/orexin A in the ventral tegmental area enhances dopamine responses to cocaine and promotes cocaine self-administration. *Psychopharmacology, 214*(2), 415–426. https://doi.org/10.1007/s00213-010-2048-8.

Espana, R. A., Oleson, E. B., Locke, J. L., Brookshire, B. R., Roberts, D. C., & Jones, S. R. (2010). The hypocretin-orexin system regulates cocaine self-administration via actions on the mesolimbic dopamine system. *The European Journal of Neuroscience*, *31*(2), 336–348. https://doi.org/10.1111/j.1460-9568.2009.07065.x.

Everitt, B. J., Dickinson, A., & Robbins, T. W. (2001). The neuropsychological basis of addictive behaviour. *Brain Research. Brain Research Reviews*, *36*(2-3), 129–138.

Gautvik, K. M., de Lecea, L., Gautvik, V. T., Danielson, P. E., Tranque, P., Dopazo, A., et al. (1996). Overview of the most prevalent hypothalamus-specific mRNAs, as identified by directional tag PCR subtraction. *Proceedings of the National Academy of Sciences of the United States of America*, *93*(16), 8733–8738.

Gotter, A. L., Webber, A. L., Coleman, P. J., Renger, J. J., & Winrow, C. J. (2012). International union of basic and clinical pharmacology. LXXXVI. Orexin receptor function, nomenclature and pharmacology. *Pharmacological Reviews*, *64*(3), 389–420. https://doi.org/10.1124/pr.111.005546.

Gozzi, A., Turrini, G., Piccoli, L., Massagrande, M., Amantini, D., Antolini, M., et al. (2011). Functional magnetic resonance imaging reveals different neural substrates for the effects of orexin-1 and orexin-2 receptor antagonists. *PLoS One*, *6*(1), e16406. https://doi.org/10.1371/journal.pone.0016406.

Harris, G. C., & Aston-Jones, G. (2003). Critical role for ventral tegmental glutamate in preference for a cocaine-conditioned environment. *Neuropsychopharmacology*, *28*(1), 73–76. https://doi.org/10.1038/sj.npp.1300011.

Harris, G. C., Wimmer, M., & Aston-Jones, G. (2005). A role for lateral hypothalamic orexin neurons in reward seeking. *Nature*, *437*(7058), 556–559. https://doi.org/10.1038/nature04071.

Haynes, A. C., Jackson, B., Chapman, H., Tadayyon, M., Johns, A., Porter, R. A., et al. (2000). A selective orexin-1 receptor antagonist reduces food consumption in male and female rats. *Regulatory Peptides*, *96*(1-2), 45–51.

Hirose, M., Egashira, S., Goto, Y., Hashihayata, T., Ohtake, N., Iwaasa, H., et al. (2003). N-acyl 6,7-dimethoxy-1,2,3,4-tetrahydroisoquinoline: the first orexin-2 receptor selective non-peptidic antagonist. *Bioorganic & Medicinal Chemistry Letters*, *13*(24), 4497–4499.

Hoch, M., Hay, J. L., Hoever, P., de Kam, M. L., te Beek, E. T., van Gerven, J. M., et al. (2013). Dual orexin receptor antagonism by almorexant does not potentiate impairing effects of alcohol in humans. *European Neuropsychopharmacology*, *23*(2), 107–117. https://doi.org/10.1016/j.euroneuro.2012.04.012.

Hollander, J. A., Pham, D., Fowler, C. D., & Kenny, P. J. (2012). Hypocretin-1 receptors regulate the reinforcing and reward-enhancing effects of cocaine: pharmacological and behavioral genetics evidence. *Frontiers in Behavioral Neuroscience*, *6*, 47. https://doi.org/10.3389/fnbeh.2012.00047.

Hutcheson, D. M., Quarta, D., Halbout, B., Rigal, A., Valerio, E., & Heidbreder, C. (2011). Orexin-1 receptor antagonist SB-334867 reduces the acquisition and expression of cocaine-conditioned reinforcement and the expression of amphetamine-conditioned reward. *Behavioural Pharmacology*, *22*(2), 173–181. https://doi.org/10.1097/FBP.0b013e328343d761.

Ito, R., Dalley, J. W., Robbins, T. W., & Everitt, B. J. (2002). Dopamine release in the dorsal striatum during cocaine-seeking behavior under the control of a drug-associated cue. *The Journal of Neuroscience*, *22*(14), 6247–6253. https://doi.org/10.1523/JNEUROSCI.22-14-06247.2002.

James, M. H., Charnley, J. L., Levi, E. M., Jones, E., Yeoh, J. W., Smith, D. W., et al. (2011). Orexin-1 receptor signalling within the ventral tegmental area, but not the paraventricular thalamus, is critical to regulating cue-induced reinstatement of cocaine-seeking. *The International Journal of Neuropsychopharmacology*, *14*(5), 684–690. https://doi.org/10.1017/S1461145711000423.

Kalivas, P. W., & O'Brien, C. (2008). Drug addiction as a pathology of staged neuroplasticity. *Neuropsychopharmacology*, *33*(1), 166–180. https://doi.org/10.1038/sj.npp.1301564.

Kalivas, P. W., & Volkow, N. D. (2005). The neural basis of addiction: a pathology of motivation and choice. *The American Journal of Psychiatry*, *162*(8), 1403–1413. https://doi.org/10.1176/appi.ajp.162.8.1403.

Kelley, A. E., & Berridge, K. C. (2002). The neuroscience of natural rewards: relevance to addictive drugs. *The Journal of Neuroscience*, *22*(9), 3306–3311. https://doi.org/10.1523/JNEUROSCI.22-09-03306.2002.

Kirouac, G. J. (2015). Placing the paraventricular nucleus of the thalamus within the brain circuits that control behavior. *Neuroscience and Biobehavioral Reviews*, *56*, 315–329. https://doi.org/10.1016/j.neubiorev.2015.08.005.

Kirouac, G. J., Parsons, M. P., & Li, S. (2005). Orexin (hypocretin) innervation of the paraventricular nucleus of the thalamus. *Brain Research*, *1059*(2), 179–188. https://doi.org/10.1016/j.brainres.2005.08.035.

Langmead, C. J., Jerman, J. C., Brough, S. J., Scott, C., Porter, R. A., & Herdon, H. J. (2004). Characterisation of the binding of [3H]-SB-674042, a novel nonpeptide antagonist, to the human orexin-1 receptor. *British Journal of Pharmacology*, *141*(2), 340–346. https://doi.org/10.1038/sj.bjp.0705610.

LeSage, M. G., Perry, J. L., Kotz, C. M., Shelley, D., & Corrigall, W. A. (2010). Nicotine self-administration in the rat: effects of hypocretin antagonists and changes in hypocretin mRNA. *Psychopharmacology*, *209*(2), 203–212. https://doi.org/10.1007/s00213-010-1792-0.

Li, J., Hu, Z., & de Lecea, L. (2014). The hypocretins/orexins: integrators of multiple physiological functions. *British Journal of Pharmacology*, *171*(2), 332–350. https://doi.org/10.1111/bph.12415.

Lin, L., Faraco, J., Li, R., Kadotani, H., Rogers, W., Lin, X., et al. (1999). The sleep disorder canine narcolepsy is caused by a mutation in the hypocretin (orexin) receptor 2 gene. *Cell*, *98*(3), 365–376.

Mahler, S. V., Smith, R. J., & Aston-Jones, G. (2013). Interactions between VTA orexin and glutamate in cue-induced reinstatement of cocaine seeking in rats. *Psychopharmacology*, *226*(4), 687–698. https://doi.org/10.1007/s00213-012-2681-5.

Mahler, S. V., Smith, R. J., Moorman, D. E., Sartor, G. C., & Aston-Jones, G. (2012). Multiple roles for orexin/hypocretin in addiction. *Progress in Brain Research*, *198*, 79–121. https://doi.org/10.1016/B978-0-444-59489-1.00007-0.

Malherbe, P., Borroni, E., Gobbi, L., Knust, H., Nettekoven, M., Pinard, E., et al. (2009). Biochemical and behavioural characterization of EMPA, a novel high-affinity, selective antagonist for the OX(2) receptor. *British Journal of Pharmacology*, *156*(8), 1326–1341. https://doi.org/10.1111/j.1476-5381.2009.00127.x.

Malherbe, P., Borroni, E., Pinard, E., Wettstein, J. G., & Knoflach, F. (2009). Biochemical and electrophysiological characterization of almorexant, a dual orexin 1 receptor (OX1)/orexin 2 receptor (OX2) antagonist: comparison with selective OX1 and OX2 antagonists. *Molecular Pharmacology*, *76*(3), 618–631. https://doi.org/10.1124/mol.109.055152.

Marcus, J. N., Aschkenasi, C. J., Lee, C. E., Chemelli, R. M., Saper, C. B., Yanagisawa, M., et al. (2001). Differential expression of orexin receptors 1 and 2 in the rat brain. *The Journal of Comparative Neurology*, *435*(1), 6–25.

Martin-Fardon, R., Cauvi, G., Kerr, T. M., & Weiss, F. (2016). Differential role of hypothalamic orexin/hypocretin neurons in reward seeking motivated by cocaine versus palatable food. *Addiction Biology*. https://doi.org/10.1111/adb.12441.

Martin-Fardon, R., & Weiss, F. (2014). Blockade of hypocretin receptor-1 preferentially prevents cocaine seeking: comparison with natural reward seeking. *Neuroreport*, *25*(7), 485–488. https://doi.org/10.1097/WNR.0000000000000120.

Matzeu, A., Kallupi, M., George, O., Schweitzer, P., & Martin-Fardon, R. (2017). Dynorphin counteracts orexin in the paraventricular nucleus of the thalamus: cellular and behavioral evidence. *Neuropsychopharmacology*, *43*(5), 1010–1020. https://doi.org/10.1038/npp.2017.250.

Matzeu, A., Kerr, T. M., Weiss, F., & Martin-Fardon, R. (2016). Orexin-A/hypocretin-1 mediates cocaine-seeking behavior in the posterior paraventricular nucleus of the thalamus via orexin/hypocretin receptor-2. *The Journal of Pharmacology and Experimental Therapeutics*, *359*(2), 273–279. https://doi.org/10.1124/jpet.116.235945.

Matzeu, A., Weiss, F., & Martin-Fardon, R. (2016). In *Administration of orexin-A in the paraventricular nucleus of the thalamus activates hypothalamic orexin neurons in animals with a history of cocaine dependence 46rd Annual meeting of society of neuroscience, 2016.* Online.

McAtee, L. C., Sutton, S. W., Rudolph, D. A., Li, X., Aluisio, L. E., Phuong, V. K., et al. (2004). Novel substituted 4-phenyl-[1,3]dioxanes: potent and selective orexin receptor 2 (OX(2)R) antagonists. *Bioorganic & Medicinal Chemistry Letters*, *14*(16), 4225–4229. https://doi.org/10.1016/j.bmcl.2004.06.032.

McFarland, K., & Kalivas, P. W. (2001). The circuitry mediating cocaine-induced reinstatement of drug-seeking behavior. *The Journal of Neuroscience*, *21*(21), 8655–8663.

Muschamp, J. W., Hollander, J. A., Thompson, J. L., Voren, G., Hassinger, L. C., Onvani, S., et al. (2014). Hypocretin (orexin) facilitates reward by attenuating the antireward effects of its cotransmitter dynorphin in ventral tegmental area. *Proceedings of the National Academy of Sciences of the United States of America*, *111*(16), E1648–E1655. https://doi.org/10.1073/pnas.1315542111.

Nambu, T., Sakurai, T., Mizukami, K., Hosoya, Y., Yanagisawa, M., & Goto, K. (1999). Distribution of orexin neurons in the adult rat brain. *Brain Research*, *827*(1-2), 243–260.

Olney, J. J., Navarro, M., & Thiele, T. E. (2015). Binge-like consumption of ethanol and other salient reinforcers is blocked by orexin-1 receptor inhibition and leads to a reduction of hypothalamic orexin immunoreactivity. *Alcoholism: Clinical and Experimental Research*, *39*(1), 21–29. https://doi.org/10.1111/acer.12591.

Parsons, M. P., Li, S., & Kirouac, G. J. (2006). The paraventricular nucleus of the thalamus as an interface between the orexin and CART peptides and the shell of the nucleus accumbens. *Synapse*, *59*(8), 480–490. https://doi.org/10.1002/syn.20264.

Peyron, C., Faraco, J., Rogers, W., Ripley, B., Overeem, S., Charnay, Y., et al. (2000). A mutation in a case of early onset narcolepsy and a generalized absence of hypocretin peptides in human narcoleptic brains. *Nature Medicine*, *6*(9), 991–997. https://doi.org/10.1038/79690.

Peyron, C., Tighe, D. K., van den Pol, A. N., de Lecea, L., Heller, H. C., Sutcliffe, J. G., et al. (1998). Neurons containing hypocretin (orexin) project to multiple neuronal systems. *The Journal of Neuroscience*, *18*(23), 9996–10015.

Porrino, L. J., Coons, E. E., & MacGregor, B. (1983). Two types of medial hypothalamic inhibition of lateral hypothalamic reward. *Brain Research*, *277*(2), 269–282.

Porter, R. A., Chan, W. N., Coulton, S., Johns, A., Hadley, M. S., Widdowson, K., et al. (2001). 1,3-Biarylureas as selective non-peptide antagonists of the orexin-1 receptor. *Bioorganic & Medicinal Chemistry Letters*, *11*(14), 1907–1910.

Prince, C. D., Rau, A. R., Yorgason, J. T., & España, R. A. (2015). Hypocretin/Orexin regulation of dopamine signaling and cocaine self-administration is mediated predominantly by hypocretin receptor 1. *ACS Chemical Neuroscience*, *6*(1), 138–146. https://doi.org/10.1021/cn500246j.

Renzulli, C., Nash, M., Wright, M., Thomas, S., Zamuner, S., Pellegatti, M., et al. (2011). Disposition and metabolism of [14C]SB-649868, an orexin 1 and 2 receptor antagonist, in humans. *Drug Metabolism and Disposition*, *39*(2), 215–227. https://doi.org/10.1124/dmd.110.035386.

Sakurai, T. (2007). The neural circuit of orexin (hypocretin): maintaining sleep and wakefulness. *Nature Reviews. Neuroscience*, *8*(3), 171–181. https://doi.org/10.1038/nrn2092.

Sakurai, T., Amemiya, A., Ishii, M., Matsuzaki, I., Chemelli, R. M., Tanaka, H., et al. (1998). Orexins and orexin receptors: a family of hypothalamic neuropeptides and G protein-coupled receptors that regulate feeding behavior. *Cell, 92*(4), 573–585.

Sakurai, T., & Mieda, M. (2011). Connectomics of orexin-producing neurons: interface of systems of emotion, energy homeostasis and arousal. *Trends in Pharmacological Sciences, 32*(8), 451–462. https://doi.org/10.1016/j.tips.2011.03.007.

Sakurai, T., Nagata, R., Yamanaka, A., Kawamura, H., Tsujino, N., Muraki, Y., et al. (2005). Input of orexin/hypocretin neurons revealed by a genetically encoded tracer in mice. *Neuron, 46*(2), 297–308. https://doi.org/10.1016/j.neuron.2005.03.010.

Scammell, T. E., & Winrow, C. J. (2011). Orexin receptors: pharmacology and therapeutic opportunities. *Annual Review of Pharmacology and Toxicology, 51*, 243–266. https://doi.org/10.1146/annurev-pharmtox-010510-100528.

Schmeichel, B. E., Herman, M. A., Roberto, M., & Koob, G. F. (2017). Hypocretin neurotransmission within the central amygdala mediates escalated cocaine self-administration and stress-induced reinstatement in rats. *Biological Psychiatry, 81*(7), 606–615. https://doi.org/10.1016/j.biopsych.2016.06.010 Epub 2016 Jun 16.

Schmeichel, B. A., Matzeu, A., Koebel, P., Vendruscolo, L. F., Kieffer, B. L., Koob, G. F., et al. (2018). Knockdown of hypocretin/orexin attenuates compulsive-like cocaine self-administration in rats. *Neuropsychopharmacology*. https://doi.org/10.1038/s41386-018-0054-4 [Epub ahead of print].

See, R. E. (2002). Neural substrates of conditioned-cued relapse to drug-seeking behavior. *Pharmacology, Biochemistry, and Behavior, 71*(3), 517–529.

Smart, D., Sabido-David, C., Brough, S. J., Jewitt, F., Johns, A., Porter, R. A., et al. (2001). SB-334867-A: the first selective orexin-1 receptor antagonist. *British Journal of Pharmacology, 132*(6), 1179–1182. https://doi.org/10.1038/sj.bjp.0703953.

Smith, R. J., See, R. E., & Aston-Jones, G. (2009). Orexin/hypocretin signaling at the orexin 1 receptor regulates cue-elicited cocaine-seeking. *The European Journal of Neuroscience, 30*(3), 493–503. https://doi.org/10.1111/j.1460-9568.2009.06844.x.

Smith, R. J., Tahsili-Fahadan, P., & Aston-Jones, G. (2010). Orexin/hypocretin is necessary for context-driven cocaine-seeking. *Neuropharmacology, 58*(1), 179–184. https://doi.org/10.1016/j.neuropharm.2009.06.042.

Steiner, M. A., Gatfield, J., Brisbare-Roch, C., Dietrich, H., Treiber, A., Jenck, F., et al. (2013). Discovery and characterization of ACT-335827, an orally available, brain penetrant orexin receptor type 1 selective antagonist. *ChemMedChem, 8*(6), 898–903. https://doi.org/10.1002/cmdc.201300003.

Steiner, M. A., Lecourt, H., & Jenck, F. (2013). The dual orexin receptor antagonist almorexant, alone and in combination with morphine, cocaine and amphetamine, on conditioned place preference and locomotor sensitization in the rat. *The International Journal of Neuropsychopharmacology, 16*(2), 417–432. https://doi.org/10.1017/S1461145712000193.

Thannickal, T. C., Moore, R. Y., Nienhuis, R., Ramanathan, L., Gulyani, S., Aldrich, M., et al. (2000). Reduced number of hypocretin neurons in human narcolepsy. *Neuron, 27*(3), 469–474.

Trivedi, P., Yu, H., MacNeil, D. J., Van der Ploeg, L. H., & Guan, X. M. (1998). Distribution of orexin receptor mRNA in the rat brain. *FEBS Letters, 438*(1-2), 71–75.

Tsujino, N., & Sakurai, T. (2013). Role of orexin in modulating arousal, feeding, and motivation. *Frontiers in Behavioral Neuroscience*, 7, 28. https://doi.org/10.3389/fnbeh.2013.00028.

Wanat, M. J., Willuhn, I., Clark, J. J., & Phillips, P. E. (2009). Phasic dopamine release in appetitive behaviors and drug addiction. *Current Drug Abuse Reviews, 2*(2), 195–213.

Wang, B., You, Z. B., & Wise, R. A. (2009). Reinstatement of cocaine seeking by hypocretin (orexin) in the ventral tegmental area: independence from the local corticotropin-releasing factor network. *Biological Psychiatry*, *65*(10), 857–862. https://doi.org/10.1016/j.biopsych.2009.01.018.

Whitman, D. B., Cox, C. D., Breslin, M. J., Brashear, K. M., Schreier, J. D., Bogusky, M. J., et al. (2009). Discovery of a potent, CNS-penetrant orexin receptor antagonist based on an n,n-disubstituted-1,4-diazepane scaffold that promotes sleep in rats. *ChemMedChem*, *4*(7), 1069–1074. https://doi.org/10.1002/cmdc.200900069.

Yeoh, J. W., James, M. H., Jobling, P., Bains, J. S., Graham, B. A., & Dayas, C. V. (2012). Cocaine potentiates excitatory drive in the perifornical/lateral hypothalamus. *The Journal of Physiology*, *590*(16), 3677–3689. https://doi.org/10.1113/jphysiol.2012.230268.

Zhou, L., Smith, R. J., Do, P. H., Aston-Jones, G., & See, R. E. (2012). Repeated orexin 1 receptor antagonism effects on cocaine seeking in rats. *Neuropharmacology*, *63*(7), 1201–1207. https://doi.org/10.1016/j.neuropharm.2012.07.044.

Zhou, L., Sun, W. L., & See, R. E. (2011). Orexin receptor targets for anti-relapse medication development in drug addiction. *Pharmaceuticals (Basel)*, *4*(6), 804–821. https://doi.org/10.3390/ph4060804.

Zhou, Y., Cui, C. L., Schlussman, S. D., Choi, J. C., Ho, A., Han, J. S., et al. (2008). Effects of cocaine place conditioning, chronic escalating-dose "binge" pattern cocaine administration and acute withdrawal on orexin/hypocretin and preprodynorphin gene expressions in lateral hypothalamus of Fischer and Sprague-Dawley rats. *Neuroscience*, *153*(4), 1225–1234. https://doi.org/10.1016/j.neuroscience.2008.03.023.

CHAPTER 7

Orexins and Cognition: Neuroanatomical and Neurochemical Substrates

Joshua A. Burk*, Eden B. Maness*, Sarah A. Blumenthal*, Jim R. Fadel†
*Department of Psychological Sciences, College of William and Mary, Williamsburg, VA, United States
†Department of Pharmacology, Physiology & Neuroscience, University of South Carolina School of Medicine, Columbia, SC, United States

The hypothalamus is the key CNS "node" for many aspects of physiological regulation, including feeding, drinking, reproduction, and thermoregulation. Neuroscientists and cognitive psychologists have long recognized that these behaviors have cognitive components. Homeostatic feeding, for example, requires an interoceptive awareness of the hunger state, attention to environmental cues that signal availability of food or the presence of threats that may be encountered during food-seeking, and spatial navigation informed by recall of prior food locations and forming of new food-related memories (Shin, Zheng, & Berthoud, 2009). Similar cognitive processes likely govern the behavioral response to multiple other homeostatic challenges for which the physiological response is coordinated at hypothalamic levels. Finally, all of these behaviors may be facilitated by heightened levels of arousal and wakefulness. At an intuitive level, then, it makes sense that the hypothalamus can modulate the activity of other forebrain or brainstem areas and circuits that mediate cognitive function. However, the specific hypothalamic cell populations, neurotransmitters, and circuit dynamics that underlie hypothalamic regulation of cognition remain to be fully elucidated.

The discovery of the orexin/hypocretin system in 1998 (de Lecea et al., 1998; Sakurai et al., 1998) spurred intense interest in the role of these peptides in a variety of physiological and behavioral functions. Because orexin neurons—particularly the lateral bank of these cells—are distributed densely in the classic "feeding center" of the brain, the lateral hypothalamus (LH), initial investigations into their functions centered largely on regulation of food intake. And indeed, early evidence suggested that orexins facilitated

The Orexin/Hypocretin System
https://doi.org/10.1016/B978-0-12-813751-2.00007-3

food intake (Sakurai et al., 1998; Sweet, Levine, Billington, & Kotz, 1999) and food-anticipatory behavior (Akiyama et al., 2004; Mieda et al., 2004). However, it soon became clear that orexins are involved in a wide variety of physiological regulatory processes, perhaps most clearly demonstrated by the finding that narcolepsy is caused by a loss of orexin neurons (Nishino, Ripley, Overeem, Lammers, & Mignot, 2000; Thannickal et al., 2000). In-depth discussions of the roles of the orexin system in feeding, stabilization of sleep-wake states, and drug addiction are found elsewhere in this volume, but to the extent that all of these have cognitive components, we will focus here on the role of the orexin system in attention, learning, and memory.

OREXINS IN LEARNING AND MEMORY

Hypothalamic orexin neurons project to multiple midbrain and forebrain areas that underlie different aspects of learning and memory, including the hippocampal formation, medial septum, amygdala, and prefrontal cortex (Peyron et al., 1998). Through these connections, orexins have been shown to contribute to and strengthen learning- and memory-related cellular processes. Orexins also facilitate global neuronal excitation through a variety of neurobiological networks, both directly and indirectly promoting the release of other cognition-supporting peptides. In particular, the widespread influence of this network is observed in cortical and subcortical structures involved in motivation-reliant processes in the context of unconditioned and conditioned behaviors, including reward acquisition, avoidance, and fear. This section focuses on incentive-based learning and memory as other chapters in this book focus on other forms of learning (e.g., fear learning).

Orexins in Incentive Learning and Memory

Orexins have been classically associated with appetitive regulation and food acquisition. In general, however, orexinergic activity is highly involved in many processes which encourage engagement with and learning about the surrounding environment. In 2014, Sakurai posited that, while orexin neurons originating in the perifornical area (PFA) and dorsomedial area of the hypothalamus are involved in circadian rhythmicity, wakefulness, and arousal, orexin-releasing cells in the LH innervate frontal and cortical brain regions and regulate reward-related learning processes (Sakurai, 2014). Despite this dichotomy in structure and function, general augmentation of orexin neurotransmission underlies many mutually supportive psychological states such as consciousness, alertness, vigilance, and motivation

(Tsujino & Sakurai, 2009). Spikes in orexin neuron firing coincide most with purposeful and motivated behaviors, including exploration, play, predation, and grooming, as well as food- and water-seeking behaviors (Mahler, Moorman, Smith, James, & Aston-Jones, 2014). Incentivized activation is an adaptive psychological state that is not only implicated in behaviors related to homeostatic regulation and environmental exploration, but learning and memory as well, especially in the context of reward acquisition. Orexin-1 (Ox1R) and orexin-2 receptors (Ox2Rs) agonists have been shown to enhance performance in tasks that result in the receipt of either a food or drug reward contingent upon accurate responding, helping to promote motivation for positive reinforcement (Harris, Wimmer, & Aston-Jones, 2005).

Orexin regulation of reinforcement and reward learning is thought to depend, in large part, on its interactions with the mesocorticolimbic dopamine (DA) system. The dopaminergic neurons of the ventral tegmental area (VTA) receive inputs from the LH which facilitate cortical and nucleus accumbens (NAcc) DA efflux (Fadel & Deutch, 2002; Korotkova, Sergeeva, Eriksson, Haas, & Brown, 2003; Trivedi, Yu, MacNeil, Van der Ploeg, & Guan, 1998). There is extensive overlap of the projection fields of orexin neurons with both dopaminergic and noradrenergic systems (Baldo, Daniel, Berridge, & Kelley, 2003; Fadel & Deutch, 2002) and VTA orexinergic transmission, in particular, encourages reinforcement-seeking behaviors (Calipari & Espana, 2012; Espana, 2012). In fact, the integrity of lateral hypothalamic innervations of the VTA appears to be necessary for the learning and execution of goal-oriented actions directed at either natural or drug reinforcers (Espana et al., 2010; Harris, Wimmer, Randall-Thompson, & Aston-Jones, 2007). The majority of VTA neurons are activated following intracranial OxA administration, resulting in elevated dopaminergic activity within both the VTA and the prefrontal cortex (PFC) (Korotkova et al., 2003; Vittoz & Berridge, 2006). Tonic OxA application to medial PFC neurons produces an increase in short- and long-latency excitatory activation of the VTA as well, particularly during the active period of the circadian rhythm (Moorman & Aston-Jones, 2010).

Much of the stimulation of VTA cells following OxA administration is believed to be a result of its influence on the activation of cortical alpha-amino-3-hydroxy-5-methyl-4-isoxazolepropionic acid receptors (AMPARs) and *N*-methyl-D-aspartate receptors (NMDARs). Orexin receptors are highly expressed throughout the hippocampus—an area rich in glutamatergic pyramidal cells and glutamate receptors—and the administration of orexins into this brain region greatly promotes synaptic plasticity (Marcus et al., 2001; Selbach et al., 2004). Because glutamate is integral for rapid excitatory signaling, its elevated release via orexin receptor input aids brain-wide activation, including in frontal structures critical to information processing. When applied to midbrain slices *in vitro*, orexins stimulate presynaptic glutamate release and enhance potentiation of postsynaptic NMDARs; similarly, both OxA and OxB isoforms regulate glutamate-mediated synaptic transmission by enhancing or lessening long-term potentiation (LTP) in the VTA, respectively (Borgland, Storm, & Bonci, 2008; Borgland, Taha, Sarti, Fields, & Bonci, 2006).

Orexins have also been shown to increase the release of cortical and hippocampal acetylcholine (ACh). Cholinergic neurotransmission in these areas is necessary for not only attentional processing, but for both the acquisition and formation of conditioned behaviors and their associated memory consolidation (Baxter & Chiba, 1999; Sarter & Bruno, 1997). Both muscarinic (mACh) and nicotinic (nACh) receptor subtypes are critical for the generation and propagation of mesolimbic and mesocortical LTP, especially in structures that are important for encoding episodic memories, such as the entorhinal cortex, the perirhinal cortex, and the hippocampus (Hasselmo, 2006). The formation of novel memories, but not necessarily the retrieval of those that were previously learned, is highly reliant on ACh levels. In support of this, the pharmacological inhibition or lesioning of cholinergic neurons precludes the formation of new memories in both rats (Bunce, Sabolek, & Chrobak, 2004; McGaughy et al., 2005) and humans (Atri et al., 2004), and their depolarization with cholinomimetic agents strengthens memory-related cellular processes (Green et al., 2005). Hypothalamic orexinergic neurons make synaptic contacts with neurons of the basal forebrain cholinergic system (BFCS)—which sends dense fibers to prefrontal cortical cells that release not only ACh, but also glutamate and gamma-aminobutyric acid (GABA) (Henny and Jones, 2008)—as well as onto cortical and subcortical structures expressing ample ACh receptors (Eggermann et al., 2001). The cholinergic ventral ascending arousal system, a pathway that enables conscious awareness through posterior hypothalamic inputs, is activated via bilateral infusions of OxA into the basal forebrain (BF) of anesthetized rats (Dong, Fukuda, Murata, Zhu, & Higuchi, 2006). Drugs which agonize ACh-releasing receptors, like nicotine, have also been found to augment the propagation of orexin neuropeptides as well as transiently heighten the acquisition and formation of memories, creating a positive feedback loop between these two transmitters (Yamanaka et al., 2003; Pasumarthi & Fadel, 2008).

OREXINS AND ATTENTION

Attention, or the detection and selection of relevant environmental stimuli and discrimination of these stimuli from noise, is controlled by cortical projections to the sensory, prefrontal, and parietal regions (Robbins, 1997). Understanding the mechanisms underlying attention furthers our insight into cognitive processes and the etiology of attention-related disorders, such as attentional deficit hyperactivity disorder (ADHD) and age-related cognitive decline. Attention requires a generalized state of physiological readiness, also referred to as arousal, to detect relevant stimuli (Robbins, 1997). Arousal and attention are closely related, but are assessed differently, where changes in arousal are detected from brain activity, such as electroencephalographic (EEG) data, and changes in attention are determined by tests of behavioral performance (Coull, 1998). Attention is dependent on a state of arousal, and orexin neurons are hypothesized to be a critical neural mechanism that maintains the relationship between arousal and attention (Fadel & Burk, 2010).

Orexin neurons originating in the LH innervate the ascending arousal system, projecting to areas implicated in wakefulness such as the locus coeruleus (LC), the BF, and the tuberomammillary nucleus (TMN) (Yoshida, McCormack, Espana, Crocker, & Scammell, 2006). Both Ox1Rs and Ox2Rs have been found in the LC, BF, and midbrain raphe nuclei and reticular formation, structures collectively critical for attention and arousal (Estabrooke et al., 2001; Frederick-Duus, Guyton, & Fadel, 2007; Marcus et al., 2001). Orexin activation maintains wakefulness through excitatory receptors which increase activity in the arousal system. Expression of the neuronal activity marker, Fos, in orexin neurons is positively correlated with wakefulness and negatively correlated with non-REM and REM sleep, providing evidence that orexin activation contributes to the maintenance of wakefulness (Estabrooke et al., 2001). Furthermore, Hagan et al. (1999) found that intraventricular OxA administration promoted wakefulness in rodents, by dose dependently increasing arousal and suppressing non-REM and REM sleep. The OxA administration also facilitated behaviors indicative of an aroused state such as increased locomotor activity, rearing, grooming, searching, and consumption of food.

The importance of orexin signaling in promoting and maintaining wakefulness is best illustrated by narcolepsy, a sleep disorder, associated with significant disruptions in orexin signaling. Narcolepsy is a debilitating neurological disease characterized by excessive daytime sleepiness, often in the form of involuntary or irresistible daytime "sleep attacks" which can occur during normal everyday activities, such as talking, eating, or driving. Soon after the discovery of the orexins, a link with sleep disorders was suggested by animal studies in which spontaneous or experimentally induced genetic deficits in orexin expression or receptor function were associated with a narcoleptic phenotype (Chemelli et al., 1999; Lin et al., 1999; Reilly, 1999; Siegel, 1999) Consistent with these animal findings, postmortem examination of the brains of human narcolepsy patients showed a marked reduction in the number of orexin neurons present as compared to controls (Peyron et al., 2000; Thannickal et al., 2000). Cerebrospinal fluid samples from narcoleptic patients also reveal dramatically lower levels of orexin (Nishino et al., 2000), and orexin deficiency, concurrent with symptoms of irresistible sleepiness, is part of the DSM-5 criteria for the disorder.

Aside from excessive sleepiness, narcoleptics often complain of attentional problems which lead to difficulties in educational and occupational performance (Rieger, Mayer, & Gauggel, 2003; Zamarian et al., 2015). Narcoleptics generally have normal cognition and can maintain arousal during tasks requiring attention and alertness for short periods of time; however, this arousal becomes more variable and worsens overall when sustained attention tasks last longer than 10 min, indicative of deficits in sustained attention. In addition, deficits in divided attention and flexible attention have been reported in patients with narcolepsy (Rieger et al., 2003). These attentional deficits occur even during periods of normal wakefulness, suggesting that this decline in attention is not due solely to sleepiness, but may be caused by inappropriate processing of relevant stimuli and is the consequence of reduced signaling to the cortex (Rieger et al., 2003). The problems with wakefulness and attention seen in narcolepsy, a disorder characterized by orexin loss, illustrate the role orexins play in not only arousal, but in attentional processes as well.

The LH has long been recognized as a source of input to the basal forebrain cholinergic system that may be important for wakefulness and attention. Orexin neurons appear to be an important component of this projection as a robust plexus of orexin fibers is seen in the basal forebrain, where they form numerous apparent appositional contacts on magnocellular cholinergic neurons (Cullinan & Zaborszky, 1991; Zaborszky & Cullinan, 1989). In addition to direct projections to the cortex, the orexin–basal forebrain projection allows for another mechanism to modulate cognitive function, as many aspects of attention are dependent on the BFCS. Higher ACh levels are associated with aroused states and heightened attentional performance (Pepeu & Giovannini, 2004). In rodent studies, increases in ACh efflux accompany increases in attentional demand, providing evidence for the relationship between attentional effort and cortical ACh release (Himmelheber, Sarter, & Bruno, 2000). In studies with humans, pharmacological stimulation of cholinergic receptors can facilitate visual and spatial attention (Bentley, Husain, & Dolan, 2004). Many studies have examined the effects of nicotine, an ACh receptor agonist, as a means of improving attentional function in both humans and primate models (Sahakian, Jones, Levy, Gray, & Warburton, 1989; Thiel, Zilles, & Fink, 2005; Witte, Davidson, & Marrocco, 1997). Furthermore, several lines of evidence demonstrate that orexin neurons are targets of acute or chronic nicotine administration, suggesting a role for the orexin system in neurochemical and cognitive correlates of nicotine (Corrigall, 2009; Lambe, Olausson, Horst, Taylor, & Aghajanian, 2005; Pasumarthi & Fadel, 2008). Damage to basal forebrain corticopetal neurons in rodent models, using the immunotoxin 192 IgG-saporin, results in severe deficits in sustained and divided attention, and poor performance on attention-demanding tasks (Himmelheber et al., 2000; McGaughy, Kaiser, & Sarter, 1996). The loss of cortical cholinergic inputs is thought to be one of the main causes of cognitive decline associated with Alzheimer's disease (AD) (Mesulam, 2013). Interestingly, postmortem studies suggest that AD patients have fewer orexin neurons (Fronczek et al., 2012). Furthermore, animal studies suggest that aging alone is associated with decreased expression of orexin peptides and receptors (Kessler, Stanley, Frederick-Duus, & Fadel, 2011; Porkka-Heiskanen, Alanko, Kalinchuk, Heiskanen, & Stenberg, 2004), suggesting a potential orexin contribution to age-related cognitive decline.

Both orexin receptor subtypes are located within the BF (Kilduff & de Lecea, 2001). These orexin receptors are distributed within the areas of the BF that include corticopetal septohippocampal cholinergic neurons, and pharmacological studies suggest that orexins can act directly within the BF to activate cholinergic neurons and promote ACh release in cortical and hippocampal regions (Eggermann et al., 2001, Fadel & Frederick-Duus, 2008, Arrigoni, Mochizuki, & Scammell, 2010, Stanley & Fadel, 2011). Some experiments have found that OxA is more potent than OxB with regard to increasing ACh efflux and EEG activity (Dong et al., 2006; Fadel, Pasumarthi, & Reznikov, 2005) suggesting a larger Ox1R contribution to orexin activation of BFCS neurons. For example, increases in prefrontal cortical ACh efflux occurring during feeding-related arousal are blocked by systemic or intrabasalis administration of the Ox1R antagonist

SB-334867 (Boschen, Fadel, & Burk, 2009; Frederick-Duus et al., 2007). Boschen and colleagues also found that intrabasalis administration of SB-334867 resulted in decreased performance on a sustained visual attention task, further supporting the contribution of Ox1Rs in attentional processing within the BFCS (Boschen et al., 2009). However, they found similar decreases in attentional performance with systemic administration of the Ox1R antagonist, SB-334867, suggesting that neurotransmitter systems outside of the BFCS may contribute to or mediate some of these effects. Furthermore, electrophysiological recordings in identified cholinergic neurons from rat brain slices have suggested primarily an Ox2R-mediated excitatory effect of orexins on these neurons (Eggermann et al., 2001).

Noncholinergic neurons in the BF are also excited by orexins. Administration of OxA increases local glutamate release within the BF, which indirectly modulates BF cholinergic activity (Frederick-Duus et al., 2007). Additionally, GABAergic neurons in the BF responsible for arousal signaling to the cortex are directly excited by OxA, suggesting the importance of orexins in modulating this pathway (Arrigoni et al., 2010). Furthermore, Blanco-Centurion and colleagues found that, following selective lesions to BF cholinergic neurons in rats, direct administration of OxA to the BF still promoted arousal, indicating that OxA acts on noncholinergic neurons in the BF to aid in arousal and associated attentional processes (Blanco-Centurion, Gerashchenko, Salin-Pascual, & Shiromani, 2004).

Systemic administration of drugs that modulate orexin transmission provides further evidence for the influence of these peptides on attention. In rodents systemic injections of SB-334867, an Ox1R antagonist, were found to disrupt performance on a sustained attention task and decrease signal detection accuracy (Boschen et al., 2009). This decrease in signal detection accuracy was not found when SB-334867 was infused directly in the BF, although both intracranial and systemic administration yielded attentional deficits. This suggests that neuronal pathways outside of the BF may mediate some of the effects of orexin on attention. One such pathway is from the paraventricular thalamic nucleus (PVT) to the cortex. The PVT receives dense innervations from orexin neurons arising in the LH, and projects to several cortical areas, including the medial prefrontal cortex (mPFC) (Huang, Ghosh, & van den Pol, 2006), a cortical region integral in attentional processes and cognition, specifically sustained attention (Sarter, Givens, & Bruno, 2001). Electrophysiological recordings and calcium imaging in rat brain slices indicate that cortically projecting glutamatergic neurons from the PVT are excited and depolarized by both OxA and

OxB; however, the effects of OxB are stronger, suggesting that orexins excite PVT neurons through Ox2R activation (Ishibashi et al., 2005). This excitation and depolarization may be mediated by activation of nonselective cation channels as well as the closure of potassium channels (Huang et al., 2006; Lambe & Aghajanian, 2003). Orexin-mediated glutamate efflux from the PVT to the mPFC leads to cortical excitation which may facilitate attentional processes arising from the mPFC; however, future studies should examine the effects of direct PVT orexin administration on attention.

Orexins may also promote attentional processes through direct signaling to the mPFC. Orexin-immunoreactive axons provide a robust innervation of the entire PFC, forming a particularly dense plexus in deeper layers of the mPFC (Fadel & Deutch, 2002). Both Ox1Rs and Ox2Rs are distributed throughout the mPFC, and the effects of direct application of orexin peptides in this area give evidence for the importance of orexin signaling on attention (Cluderay, Harrison, & Hervieu, 2002; Hervieu, Cluderay, Harrison, Roberts, & Leslie, 2001; Jin et al., 2016). In mPFC pyramidal neurons, OxA application leads to excitation by closing potassium channels and activating extracellular calcium influx (Xia et al., 2005). This excitatory effect of OxA has been found throughout multiple layers within the mPFC (Li et al., 2010; Yan, He, Xia, Zhang, & Hu, 2012). The Ox1R antagonist, SB-334867, was found to block the excitatory effects of OxA in layer 5 of the mPFC, suggesting an Ox1R specific role in promoting this excitation (Li et al., 2010). In rodents, OxB infusions directly into the mPFC led to significant improvements in performance on a sustained and divided attention task with high attentional demand (Lambe et al., 2005). These studies suggest that orexins may regulate mPFC signaling through direct innervation to mPFC neurons, and thereby may have a direct effect on mPFC-directed attentional processes. However, the direct effects of orexins in the mPFC should be studied more, to examine the exact roles of OxA and OxB and the orexin receptor subtypes on attention. Furthermore, ACh-orexin interactions within the mPFC should be examined, to understand the role of these interactions on attention.

Studies on the potential clinical efficacy of orexin-based treatments in humans bring further insight into the importance of orexins in attention. Currently, orexin-based drugs are being examined as potential treatment options for those suffering from narcolepsy (Nishino et al., 2000; Peyron et al., 2000). Given the relative paucity of nonpeptide orexin agonists and the limited capacity of peripherally administered orexins to access the brain, direct intranasal administration of OxA has been suggested as an effective means of rapidly targeting brain regions implicated in wakefulness and attention (Dhuria, Hanson, & Frey 2nd, 2009). Intranasal OxA activates brain regions and neuronal populations implicated in cognitive function and stimulates prefrontal cortical ACh release (Calva, Fayyaz, & Fadel, 2018). In clinical studies following intranasal administration of OxA, narcoleptic patients showed enhanced performance on a divided attention task, measured by reaction time and number of false reactions (Weinhold et al., 2014). This task was particularly sensitive to attentional impairments due to tiredness and provides evidence for the attention-enhancing effects of OxA in humans. In nonhuman primates, intranasal administration of OxA was also effective at improving cognitive function in sleep-deprived animals, more so than intravenous administration (Deadwyler, Porrino, Siegel, & Hampson, 2007). Intranasal administration of orexins may not only be an effective treatment option for those suffering from narcolepsy, but also for others suffering from other attentional deficits, such as those seen in the cognitive decline associated with age-related disorders, due to the critical role orexins play in attention.

CONCLUSIONS

Orexin neurons have widespread connections with brain regions and neuromodulatory systems that underlie learning, memory, and attentional function. The involvement of orexins in cognitive function in both humans and animal models is suggested by the deficits associated with compromised orexin neurotransmission, whether occurring in the context of disease (e.g., narcolepsy), natural aging, or genetic or pharmacological manipulations. Collectively, these data suggest that the orexin system may be a productive target for the development of novel therapeutics for conditions in which the orexin system is implicated.

REFERENCES

Akiyama, M., Yuasa, T., Hayasaka, N., Horikawa, K., Sakurai, T., & Shibata, S. (2004). Reduced food anticipatory activity in genetically orexin (hypocretin) neuron-ablated mice. *The European Journal of Neuroscience*, *20*, 3054–3062.

Arrigoni, E., Mochizuki, T., & Scammell, T. E. (2010). Activation of the basal forebrain by the orexin/hypocretin neurones. *Acta Physiologica (Oxford, England)*, *198*, 223–235.

Atri, A., et al. (2004). Blockade of central cholinergic receptors impairs new learning and increases proactive interference in a word paired-associate memory task. *Behavioral Neuroscience*, *118*(1), 223–236.

Baldo, B. A., Daniel, R. A., Berridge, C. W., & Kelley, A. E. (2003). Overlapping distributions of orexin/hypocretin- and dopamine-beta-hydroxylase immunoreactive fibers in rat brain regions mediating arousal, motivation, and stress. *The Journal of Comparative Neurology*, *464*, 220–237.

Baxter, M. G., & Chiba, A. A. (1999). Cognitive functions of the basal forebrain. *Current Opinion in Neurobiology*, *9*, 178–183.

Bentley, P., Husain, M., & Dolan, R. J. (2004). Effects of cholinergic enhancement on visual stimulation, spatial attention, and spatial working memory. *Neuron*, *41*, 969–982.

Blanco-Centurion, C., Gerashchenko, D., Salin-Pascual, R. J., & Shiromani, P. J. (2004). Effects of hypocretin2-saporin and antidopamine-beta-hydroxylase-saporin neurotoxic lesions of the dorsolateral pons on sleep and muscle tone. *The European Journal of Neuroscience*, *19*, 2741–2752.

Borgland, S. L., Storm, E., & Bonci, A. (2008). Orexin B/hypocretin 2 increases glutamatergic transmission to ventral tegmental area neurons. *The European Journal of Neuroscience*, *28*, 1545–1556.

Borgland, S. L., Taha, S. A., Sarti, F., Fields, H. L., & Bonci, A. (2006). Orexin A in the VTA is critical for the induction of synaptic plasticity and behavioral sensitization to cocaine. *Neuron*, *49*, 589–601.

Boschen, K. E., Fadel, J. R., & Burk, J. A. (2009). Systemic and intrabasalis administration of the orexin-1 receptor antagonist, SB-334867, disrupts attentional performance in rats. *Psychopharmacology*, *206*, 205–213.

Bunce, J. G., Sabolek, H. R., & Chrobak, J. J. (2004). Timing of administration mediates the memory effects of intraseptal carbachol infusion. *Neuroscience*, *127*(3), 593–600.

Calipari, E. S., & Espana, R. A. (2012). Hypocretin/orexin regulation of dopamine signaling: implications for reward and reinforcement mechanisms. *Frontiers in Behavioral Neuroscience*, *6*, 54.

Calva, C. B., Fayyaz, H., & Fadel, J. R. (2018). Increased acetylcholine and glutamate efflux in the prefrontal cortex following intranasal orexin-A (hypocretin-1). *Journal of Neurochemistry*, *145*, 232–244.

Chemelli, R. M., Willie, J. T., Sinton, C. M., Elmquist, J. K., Scammell, T., Lee, C., et al. (1999). Narcolepsy in orexin knockout mice: molecular genetics of sleep regulation. *Cell*, *98*, 437–451.

Cluderay, J. E., Harrison, D. C., & Hervieu, G. J. (2002). Protein distribution of the orexin-2 receptor in the rat central nervous system. *Regulatory Peptides*, *104*, 131–144.

Corrigall, W. A. (2009). Hypocretin mechanisms in nicotine addiction: evidence and speculation. *Psychopharmacology*, *206*, 23–37.

Coull, J. T. (1998). Neural correlates of attention and arousal: insights from electrophysiology, functional neuroimaging and psychopharmacology. *Progress in Neurobiology*, *55*, 343–361.

Cullinan, W. E., & Zaborszky, L. (1991). Organization of ascending hypothalamic projections to the rostral forebrain with special reference to the innervation of cholinergic projection neurons. *The Journal of Comparative Neurology*, *306*, 631–667.

de Lecea, L., Kilduff, T. S., Peyron, C., Gao, X., Foye, P. E., Danielson, P. E., et al. (1998). The hypocretins: hypothalamus-specific peptides with neuroexcitatory activity. *Proceedings of the National Academy of Sciences of the United States of America*, *95*, 322–327.

Deadwyler, S. A., Porrino, L., Siegel, J. M., & Hampson, R. E. (2007). Systemic and nasal delivery of orexin-A (Hypocretin-1) reduces the effects of sleep deprivation on cognitive performance in nonhuman primates. *The Journal of Neuroscience*, *27*, 14239–14247.

Dhuria, S. V., Hanson, L. R., & Frey, W. H., 2nd. (2009). Intranasal drug targeting of hypocretin-1 (orexin-A) to the central nervous system. *Journal of Pharmaceutical Sciences*, *98*, 2501–2515.

Dong, H. L., Fukuda, S., Murata, E., Zhu, Z., & Higuchi, T. (2006). Orexins increase cortical acetylcholine release and electroencephalographic activation through orexin-1 receptor in the rat basal forebrain during isoflurane anesthesia. *Anesthesiology, 104*, 1023–1032.

Eggermann, E., Serafin, M., Bayer, L., Machard, D., Saint-Mleux, B., Jones, B. E., et al. (2001). Orexins/hypocretins excite basal forebrain cholinergic neurones. *Neuroscience, 108*, 177–181.

Espana, R. A. (2012). Hypocretin/orexin involvement in reward and reinforcement. *Vitamins and Hormones, 89*, 185–208.

Espana, R. A., Oleson, E. B., Locke, J. L., Brookshire, B. R., Roberts, D. C., & Jones, S. R. (2010). The hypocretin-orexin system regulates cocaine self-administration via actions on the mesolimbic dopamine system. *The European Journal of Neuroscience, 31*, 336–348.

Estabrooke, I. V., McCarthy, M. T., Ko, E., Chou, T. C., Chemelli, R. M., Yanagisawa, M., et al. (2001). Fos expression in orexin neurons varies with behavioral state. *The Journal of Neuroscience, 21*, 1656–1662.

Fadel, J., & Burk, J. A. (2010). Orexin/hypocretin modulation of the basal forebrain cholinergic system: role in attention. *Brain Research, 1314*, 112–123.

Fadel, J., & Deutch, A. Y. (2002). Anatomical substrates of orexin-dopamine interactions: lateral hypothalamic projections to the ventral tegmental area. *Neuroscience, 111*, 379–387.

Fadel, J., & Frederick-Duus, D. (2008). Orexin/hypocretin modulation of the basal forebrain cholinergic system: insights from in vivo microdialysis studies. *Pharmacology, Biochemistry, and Behavior, 90*, 156–162.

Fadel, J., Pasumarthi, R., & Reznikov, L. R. (2005). Stimulation of cortical acetylcholine release by orexin A. *Neuroscience, 130*, 541–547.

Frederick-Duus, D., Guyton, M. F., & Fadel, J. (2007). Food-elicited increases in cortical acetylcholine release require orexin transmission. *Neuroscience, 149*, 499–507.

Fronczek, R., van Geest, S., Frolich, M., Overeem, S., Roelandse, F. W., Lammers, G. J., et al. (2012). Hypocretin (orexin) loss in Alzheimer's disease. *Neurobiology of Aging, 33*, 1642–1650.

Green, A., et al. (2005). Muscarinic and nicotinic receptor modulation of object and spatial n-back working memory in humans. *Pharmacology Biochemistry and Behavior, 81*(3), 575–584.

Hagan, J. J., Leslie, R. A., Patel, S., Evans, M. L., Wattam, T. A., Holmes, S., et al. (1999). Orexin A activates locus coeruleus cell firing and increases arousal in the rat. *Proceedings of the National Academy of Sciences of the United States of America, 96*, 10911–10916.

Harris, G. C., Wimmer, M., & Aston-Jones, G. (2005). A role for lateral hypothalamic orexin neurons in reward seeking. *Nature, 437*, 556–559.

Harris, G. C., Wimmer, M., Randall-Thompson, J. F., & Aston-Jones, G. (2007). Lateral hypothalamic orexin neurons are critically involved in learning to associate an environment with morphine reward. *Behavioural Brain Research, 183*, 43–51.

Hasselmo, M. E. (2006). The role of acetylcholine in learning and memory. *Current Opinion in Neurobiology, 16*, 710–715.

Henny, P., & Jones, B. E. (2008). Projections from basal forebrain to prefrontal cortex comprise cholinergic, GABAergic and glutamatergic inputs to pyramidal cells or interneurons. *European Journal of Neuroscience, 27*(3), 654–670.

Hervieu, G. J., Cluderay, J. E., Harrison, D. C., Roberts, J. C., & Leslie, R. A. (2001). Gene expression and protein distribution of the orexin-1 receptor in the rat brain and spinal cord. *Neuroscience, 103*, 777–797.

Himmelheber, A. M., Sarter, M., & Bruno, J. P. (2000). Increases in cortical acetylcholine release during sustained attention performance in rats. *Brain Research. Cognitive Brain Research, 9*, 313–325.

Huang, H., Ghosh, P., & van den Pol, A. N. (2006). Prefrontal cortex-projecting glutamatergic thalamic paraventricular nucleus-excited by hypocretin: a feedforward circuit that may enhance cognitive arousal. *Journal of Neurophysiology, 95*, 1656–1668.

Ishibashi, M., Takano, S., Yanagida, H., Takatsuna, M., Nakajima, K., Oomura, Y., et al. (2005). Effects of orexins/hypocretins on neuronal activity in the paraventricular nucleus of the thalamus in rats in vitro. *Peptides, 26*, 471–481.

Jin, J., Chen, Q., Qiao, Q., Yang, L., Xiong, J., Xia, J., et al. (2016). Orexin neurons in the lateral hypothalamus project to the medial prefrontal cortex with a rostro-caudal gradient. *Neuroscience Letters, 621*, 9–14.

Kessler, B. A., Stanley, E. M., Frederick-Duus, D., & Fadel, J. (2011). Age-related loss of orexin/hypocretin neurons. *Neuroscience, 178*, 82–88.

Kilduff, T. S., & de Lecea, L. (2001). Mapping of the mRNAs for the hypocretin/orexin and melanin-concentrating hormone receptors: networks of overlapping peptide systems. *The Journal of Comparative Neurology, 435*, 1–5.

Korotkova, T. M., Sergeeva, O. A., Eriksson, K. S., Haas, H. L., & Brown, R. E. (2003). Excitation of ventral tegmental area dopaminergic and nondopaminergic neurons by orexins/hypocretins. *The Journal of Neuroscience, 23*, 7–11.

Lambe, E. K., & Aghajanian, G. K. (2003). Hypocretin (orexin) induces calcium transients in single spines postsynaptic to identified thalamocortical boutons in prefrontal slice. *Neuron, 40*, 139–150.

Lambe, E. K., Olausson, P., Horst, N. K., Taylor, J. R., & Aghajanian, G. K. (2005). Hypocretin and nicotine excite the same thalamocortical synapses in prefrontal cortex: correlation with improved attention in rat. *The Journal of Neuroscience, 25*, 5225–5229.

Li, B., Chen, F., Ye, J., Chen, X., Yan, J., Li, Y., et al. (2010). The modulation of orexin A on HCN currents of pyramidal neurons in mouse prelimbic cortex. *Cerebral Cortex, 20*, 1756–1767.

Lin, L., Faraco, J., Li, R., Kadotani, H., Rogers, W., Lin, X., et al. (1999). The sleep disorder canine narcolepsy is caused by a mutation in the hypocretin (orexin) receptor 2 gene. *Cell, 98*, 365–376.

Mahler, S. V., Moorman, D. E., Smith, R. J., James, M. H., & Aston-Jones, G. (2014). Motivational activation: a unifying hypothesis of orexin/hypocretin function. *Nature Neuroscience, 17*, 1298–1303.

Marcus, J. N., Aschkenasi, C. J., Lee, C. E., Chemelli, R. M., Saper, C. B., Yanagisawa, M., et al. (2001). Differential expression of orexin receptors 1 and 2 in the rat brain. *The Journal of Comparative Neurology, 435*, 6–25.

McGaughy, J., Kaiser, T., & Sarter, M. (1996). Behavioral vigilance following infusions of 192 IgG-saporin into the basal forebrain: selectivity of the behavioral impairment and relation to cortical AChE-positive fiber density. *Behavioral Neuroscience, 110*, 247–265.

McGaughy, J., et al. (2005). Cholinergic deafferentation of the entorhinal cortex in rats impairs encoding of novel but not familiar stimuli in a delayed nonmatch-to-sample task. *Journal of Neuroscience, 25*(44), 10273–10281.

Mesulam, M. M. (2013). Cholinergic circuitry of the human nucleus basalis and its fate in Alzheimer's disease. *The Journal of Comparative Neurology, 521*, 4124–4144.

Mieda, M., Williams, S. C., Sinton, C. M., Richardson, J. A., Sakurai, T., & Yanagisawa, M. (2004). Orexin neurons function in an efferent pathway of a food-entrainable circadian oscillator in eliciting food-anticipatory activity and wakefulness. *The Journal of Neuroscience, 24*, 10493–10501.

Moorman, D. E., & Aston-Jones, G. (2010). Orexin/hypocretin modulates response of ventral tegmental dopamine neurons to prefrontal activation: diurnal influences. *The Journal of Neuroscience, 30*, 15585–15599.

Nishino, S., Ripley, B., Overeem, S., Lammers, G. J., & Mignot, E. (2000). Hypocretin (orexin) deficiency in human narcolepsy. *Lancet, 355*, 39–40.

Pasumarthi, R. K., & Fadel, J. (2008). Activation of orexin/hypocretin projections to basal forebrain and paraventricular thalamus by acute nicotine. *Brain Research Bulletin, 77*, 367–373.

Pepeu, G., & Giovannini, M. G. (2004). Changes in acetylcholine extracellular levels during cognitive processes. *Learning & Memory*, *11*, 21–27.

Peyron, C., Faraco, J., Rogers, W., Ripley, B., Overeem, S., Charnay, Y., et al. (2000). A mutation in a case of early onset narcolepsy and a generalized absence of hypocretin peptides in human narcoleptic brains. *Nature Medicine*, *6*, 991–997.

Peyron, C., Tighe, D. K., van den Pol, A. N., de Lecea, L., Heller, H. C., Sutcliffe, J. G., et al. (1998). Neurons containing hypocretin (orexin) project to multiple neuronal systems. *The Journal of Neuroscience*, *18*, 9996–10015.

Porkka-Heiskanen, T., Alanko, L., Kalinchuk, A., Heiskanen, S., & Stenberg, D. (2004). The effect of age on prepro-orexin gene expression and contents of orexin A and B in the rat brain. *Neurobiology of Aging*, *25*, 231–238.

Reilly, C. E. (1999). I. Mutation in the hypocretin (orexin) receptor 2 gene causes canine narcolepsy. *Journal of Neurology*, *246*, 985–986.

Rieger, M., Mayer, G., & Gauggel, S. (2003). Attention deficits in patients with narcolepsy. *Sleep*, *26*, 36–43.

Robbins, T. W. (1997). Arousal systems and attentional processes. *Biological Psychology*, *45*, 57–71.

Sahakian, B., Jones, G., Levy, R., Gray, J., & Warburton, D. (1989). The effects of nicotine on attention, information processing, and short-term memory in patients with dementia of the Alzheimer type. *The British Journal of Psychiatry*, *154*, 797–800.

Sakurai, T. (2014). The role of orexin in motivated behaviours. *Nature Reviews Neuroscience*, *15*, 719–731.

Sakurai, T., Amemiya, A., Ishii, M., Matsuzaki, I., Chemelli, R. M., Tanaka, H., et al. (1998). Orexins and orexin receptors: a family of hypothalamic neuropeptides and G protein-coupled receptors that regulate feeding behavior. *Cell*, *92*, 573–585.

Sarter, M., & Bruno, J. P. (1997). Cognitive functions of cortical acetylcholine: toward a unifying hypothesis. *Brain Research Reviews*, *23*, 28–46.

Sarter, M., Givens, B., & Bruno, J. P. (2001). The cognitive neuroscience of sustained attention: where top-down meets bottom-up. *Brain Research Reviews*, *35*, 146–160.

Selbach, O., Doreulee, N., Bohla, C., Eriksson, K. S., Sergeeva, O. A., Poelchen, W., et al. (2004). Orexins/hypocretins cause sharp wave- and theta-related synaptic plasticity in the hippocampus via glutamatergic, gabaergic, noradrenergic, and cholinergic signaling. *Neuroscience*, *127*, 519–528.

Shin, A. C., Zheng, H., & Berthoud, H. R. (2009). An expanded view of energy homeostasis: neural integration of metabolic, cognitive, and emotional drives to eat. *Physiology & Behavior*, *97*, 572–580.

Siegel, J. M. (1999). Narcolepsy: a key role for hypocretins (orexins). *Cell*, *98*, 409–412.

Stanley, E. M., & Fadel, J. R. (2011). Aging-related alterations in orexin/hypocretin modulation of septo-hippocampal amino acid neurotransmission. *Neuroscience*.

Sweet, D. C., Levine, A. S., Billington, C. J., & Kotz, C. M. (1999). Feeding response to central orexins. *Brain Research*, *821*, 535–538.

Thannickal, T. C., Moore, R. Y., Nienhuis, R., Ramanathan, L., Gulyani, S., Aldrich, M., et al. (2000). Reduced number of hypocretin neurons in human narcolepsy. *Neuron*, *27*, 469–474.

Thiel, C. M., Zilles, K., & Fink, G. R. (2005). Nicotine modulates reorienting of visuospatial attention and neural activity in human parietal cortex. *Neuropsychopharmacology*, *30*, 810–820.

Trivedi, P., Yu, H., MacNeil, D. J., Van der Ploeg, L. H., & Guan, X. M. (1998). Distribution of orexin receptor mRNA in the rat brain. *FEBS Letters*, *438*, 71–75.

Tsujino, N., & Sakurai, T. (2009). Orexin/hypocretin: a neuropeptide at the interface of sleep, energy homeostasis, and reward system. *Pharmacological Reviews*, *61*, 162–176.

Vittoz, N. M., & Berridge, C. W. (2006). Hypocretin/orexin selectively increases dopamine efflux within the prefrontal cortex: involvement of the ventral tegmental area. *Neuropsychopharmacology*, *31*, 384–395.

Weinhold, S. L., Seeck-Hirschner, M., Nowak, A., Hallschmid, M., Goder, R., & Baier, P. C. (2014). The effect of intranasal orexin-A (hypocretin-1) on sleep, wakefulness and attention in narcolepsy with cataplexy. *Behavioural Brain Research*, *262*, 8–13.

Witte, E. A., Davidson, M. C., & Marrocco, R. T. (1997). Effects of altering brain cholinergic activity on covert orienting of attention: comparison of monkey and human performance. *Psychopharmacology*, *132*, 324–334.

Xia, J., Chen, X., Song, C., Ye, J., Yu, Z., & Hu, Z. (2005). Postsynaptic excitation of prefrontal cortical pyramidal neurons by hypocretin-1/orexin a through the inhibition of potassium currents. *Journal of Neuroscience Research*, *82*, 729–736.

Yamanaka, A., et al. (2003). Regulation of orexin neurons by the monoaminergic and cholinergic systems. *Biochemical and Biophysical Research Communications*, *303*(1), 120–129.

Yan, J., He, C., Xia, J. X., Zhang, D., & Hu, Z. A. (2012). Orexin-A excites pyramidal neurons in layer 2/3 of the rat prefrontal cortex. *Neuroscience Letters*, *520*, 92–97.

Yoshida, K., McCormack, S., Espana, R. A., Crocker, A., & Scammell, T. E. (2006). Afferents to the orexin neurons of the rat brain. *The Journal of Comparative Neurology*, *494*, 845–861.

Zaborszky, L., & Cullinan, W. E. (1989). Hypothalamic axons terminate on forebrain cholinergic neurons: an ultrastructural double-labeling study using PHA-L tracing and ChAT immunocytochemistry. *Brain Research*, *479*, 177–184.

Zamarian, L., Hogl, B., Delazer, M., Hingerl, K., Gabelia, D., Mitterling, T., et al. (2015). Subjective deficits of attention, cognition and depression in patients with narcolepsy. *Sleep Medicine*, *16*, 45–51.

CHAPTER 8

The Hypocretin/Orexin System and Fear Learning

África Flores*, Fernando Berrendero[†,‡]
[*]Institute of Neurosciences, Autonomous University of Barcelona, Barcelona, Spain
[†]Laboratory of Neuropharmacology, Department of Experimental and Health Sciences, Pompeu Fabra University, Barcelona, Spain
[‡]Faculty of Experimental Sciences, Francisco de Vitoria University, Madrid, Spain

THE HYPOCRETIN/OREXIN SYSTEM: MORE THAN A HOMEOSTATIC MODULATOR

It has been twenty years since the orexinergic system was simultaneously discovered by two independent laboratories (de Lecea et al., 1998; Sakurai et al., 1998). Initially thought to promote wakefulness and mediate feeding behavior (Sakurai et al., 1998), this highly conserved neuromodulator system has been demonstrated to modulate certain cognitive functions and emotional behaviors as well (Li, Hu, & de Lecea, 2014). Indeed, orexins have been hypothesized to primarily mediate behavior under situations of high motivational relevance, such as during physiological need states, exposure to threats, or reward opportunities (Mahler, Moorman, Smith, James, & Aston-Jones, 2014). Yet, the orexin system has attracted even more attention during the last years, since its dysfunction has been reported to contribute to numerous neuropsychiatric disorders, including drug addiction and anxiety disorders (Chen, de Lecea, Hu, & Gao, 2015).

The orexin system comprises two endogenous neuropeptides, orexin-A/hypocretin-1 and orexin-B/hypocretin-2, and their associated G-protein-coupled receptors, orexin/hypocretin receptor-1 (OX1R/Hcrtr-1) and orexin/hypocretin receptor-2 (OX2R/Hcrtr-2) (de Lecea et al., 1998; Sakurai et al., 1998). These receptors differ in their ligand binding affinities: while OX2R presents similar affinity for both orexin peptides, OX1R preferentially binds to orexin-A (Sakurai et al., 1998). In neurons, orexin receptor stimulation has mainly a neuroexcitatory effect. The transduction mechanism responsible for this neuroexcitation involves the activation of Gq proteins and subsequent stimulation of the phospholipase C-protein kinase C (PLC-PKC)

The Orexin/Hypocretin System
https://doi.org/10.1016/B978-0-12-813751-2.00008-5

pathway, which culminates with an enhancement of intracellular Ca2+ concentrations (Kukkonen & Leonard, 2014). Other signal transduction mechanisms may be also triggered upon orexin receptor stimulation, including the modulation of adenylyl cyclases and the production of lipidic messengers like arachidonic acid and certain endocannabinoids (Kukkonen & Leonard, 2014).

In the CNS, orexin-expressing neurons constitute a relatively small population of cells located in certain portions of the hypothalamus, including the lateral, perifornical, and dorsomedial subregions (Peyron et al., 1998). Despite small in number, orexin-containing neurons have extensive projections throughout the entire CNS. Accordingly, OX1R and OX2R are expressed broadly within the brain, showing a partially overlapping distribution in diverse regions, although some areas have a selective expression of these receptors (Marcus et al., 2001). The ubiquity of the orexin system reflects the variety of physiological functions controlled by this neuromodulator system, including the modulation of arousal and sleep-wake cycles, appetite and energy homeostasis, reward-seeking and reinforcement, and stress responsiveness (Li, Dong, Li, & Kirouac, 2014; Li, Hu, & de Lecea, 2014; Mahler et al., 2014). Recent reports revealed a role of orexins in the regulation of aversive learning and emotional memory (Flores, Saravia, Maldonado, & Berrendero, 2015), key processes that lie at the root of many fear and anxiety disorders.

FEAR LEARNING AND EXTINCTION: RELEVANCE AND MODELING IN ANIMALS

Fear is an unpleasant emotion that emerges in presence of a real or perceived imminent threat, which involves the activation of a variety of alarm responses, including autonomic, neuroendocrine, or behavioral. These responses define the state of fear that emerges from the activation of the neural defense circuits and its reflexive autonomic and somatic outputs. Transient, adaptive fear responses allow the individual to cope with dangerous or stressful situations, and therefore have a pivotal role in survival. Importantly, fear responses are plastic, which allows adjustments to the responses to the situational demands either by increasing or by decreasing these responses. Thus if fear responses to threatening stimuli are out of proportion and persist beyond the adaptive level, these emotional states can become pathological, leading to the development of an anxiety disorder (Kindt, 2014).

Classical (or Pavlovian) fear conditioning is probably the most common paradigm employed to explore fear learning and memory, both in animals and in humans. This procedure involves an associative learning process in which a neutral conditional stimulus (CS) (i.e., a light, a tone, or a setting), is repeatedly paired with an aversive unconditional stimulus (US) (i.e., an electric footshock) (Fig. 1A). After repeated pairings, reexposure to the CS alone will elicit autonomic and behavioral fear-conditioned responses on its own. In rodents, exposure to threatening stimuli elicits diverse defensive behaviors (including fight, autonomic activation, and ultrasonic vocalizations). "Freezing," a defensive behavioral response defined by the absence of movement other than breathing, is probably the most widely used measure of fear in rodent research, and is a species-specific defense response. Fear-potentiated startle is another commonly used index of fear and consists in the increase of startle response (elicited by the presentation of an intense, unexpected stimulus) when the fear-associated stimulus (the CS associated with shock) is present compared with when it is absent. This response is conserved across many species (Lang, Davis, & Ohman, 2000): in humans, eyeblink is taken as the index of the startle response, while in rodents the whole body jump is usually taken as the magnitude of startle. Other widely used measures include avoidance behavior and changes in heart rate or arterial pressure (Ganella & Kim, 2014). According to the nature of the CS, fear conditioning can be referred to as "cued," where the CS is an elemental cue (e.g., a tone, a light, an odor), or "contextual," where the CS entails the integration of different elemental cues into a general environmental representation (Riebe, Pamplona, Pamplona, Kamprath, & Wotjak, 2012). Pharmacological studies of conditioned fear typically involve three distinctive phases: (1) the acquisition phase, which refers to the actual associative phase during which paired presentations of the CS and the US are given to the animal; (2) the consolidation phase, immediately after the acquisition phase, comprises the time window required for the working/short-term memory to become a stable, long-term memory; and (3) the expression or retrieval phase, when the animal is tested with CS-alone trials to examine whether it can elicit a fear-conditioned response (VanElzakker, Dahlgren, Davis, Dubois, & Shin, 2014).

The ability to recognize the absence and/or disappearance of a threat is necessary to successfully adapt the individual's responses to the new unthreatening situation. In rodents, the most typical intervention employed to relieve fear is extinction training (Fig. 1B). Extinction training procedures are based on repeated exposures to the conditioned cue/context in the absence of the predicted aversive stimulus to reduce fear responses from exposure to exposure. The decrease in the conditioned response after

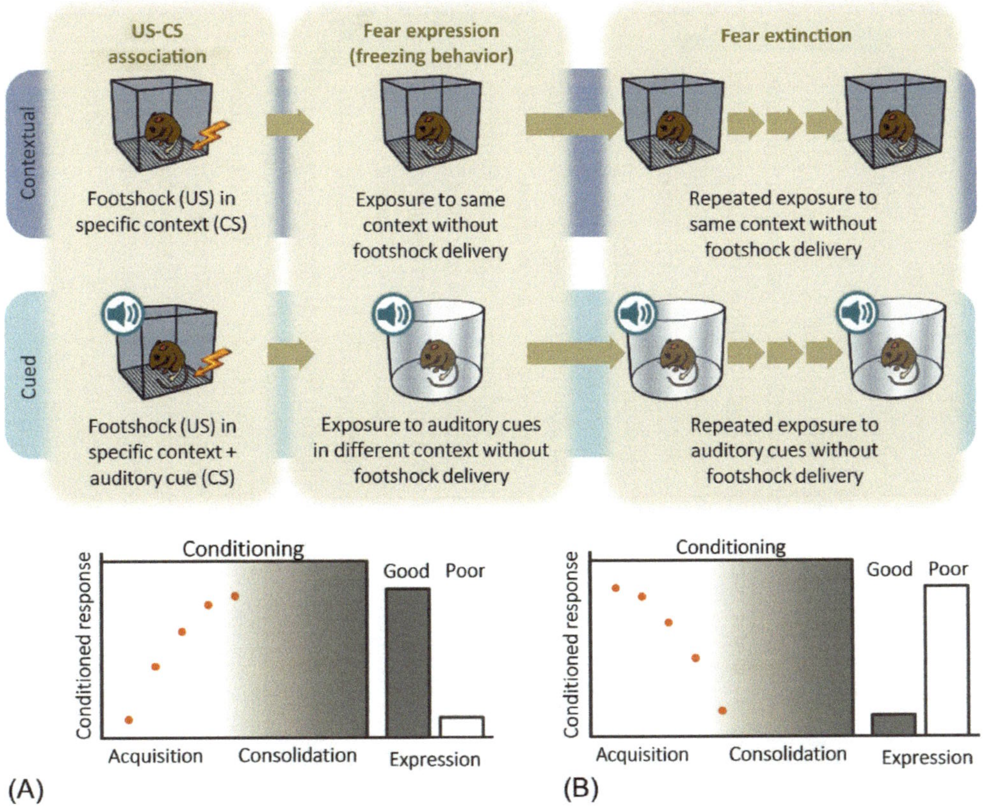

Fig. 1 Contextual and cued fear conditioning (FC) and extinction paradigms. (A) During conditioning, a particularly neutral conditioned stimulus (CS), usually a chamber in contextual FC *(upper diagram)* or a tone in cued FC *(lower diagram)*, is presented together with an aversive unconditioned stimulus (US), usually an electrical footshock. As a consequence of this US-CS association, a new exposure to the CS in absence of the US evokes a measurable conditioned response (i.e., freezing). (B) Repeated exposure to the CS in absence of the US progressively leads to fear extinction, expressed as the decrease of the freezing response and weakening of the US-CS association. Both conditioning and extinction processes comprise the phases of acquisition (during the learning session), consolidation (after the learning session), and expression (during the test session). *(Based on Ganella and Kim (2014). Modified with permission from Flores et al. (2015).)*

extinction is mainly due to new inhibitory learning (i.e., the formation of a new memory trace, which inhibits the expression of the original fear memory) (Herry et al., 2008). Extinction has received considerable attention over the past decade because of its theoretical importance and its obvious clinical implications for the treatment of various anxiety disorders (Quirk & Mueller, 2008). Indeed, the current cognitive-behavioral treatments for anxiety disorders in humans largely involve some type of exposure therapies that rely on the process of extinction (Hofmann, 2008).

BRAIN CIRCUITRY OF FEAR EXPRESSION AND EXTINCTION

The brain networks underlying fear processing are complex, and emerging studies using sophisticated techniques suggest that its functioning involves even more structures than those classically assumed. However, the primary brain core mediating fear conditioning has been well characterized in rodents and humans. A significant number of studies have identified the amygdala, the hippocampus, and the medial PFC as the classical regions of the fear neural circuit (Fig. 2), although other brain areas are also involved, including the bed nucleus of stria terminalis, the periaqueductal gray, and the thalamic paraventricular nucleus (PVT).

The amygdala is the preeminent brain structure controlling fear learning and expression, as it receives and integrates information from conditioned and unconditioned stimuli, and subsequently promotes and orchestrates a series of fear responses through projections to the behavioral, autonomic, and endocrine processing systems located in the brainstem (LeDoux, 2000). The amygdalar complex is composed of diverse nuclei, each of which executes a particular function in fear processing (LeDoux, 2007). The basolateral amygdala, constituted by the lateral amygdala (LA) and the basal amygdala (BA), is where information regarding the CS and the US converge, and represents an important site for the plastic changes underlying CS-US association (Herry & Johansen, 2014). Thus auditory and other sensory inputs from elementary cues reach the LA from both thalamic and cortical projections, and contextual information is transmitted to the BA from the hippocampus, which is responsible at the same time for assembling the different elementary contextual cues into a single representation of the context (LeDoux, 2000, 2007). US information (i.e., pain) travels from the spinal dorsal horn to the LA, probably through diverse supraspinal processing stations, including the periaqueductal gray and the thalamus (Herry & Johansen, 2014). On the other hand, the central nucleus of the amygdala

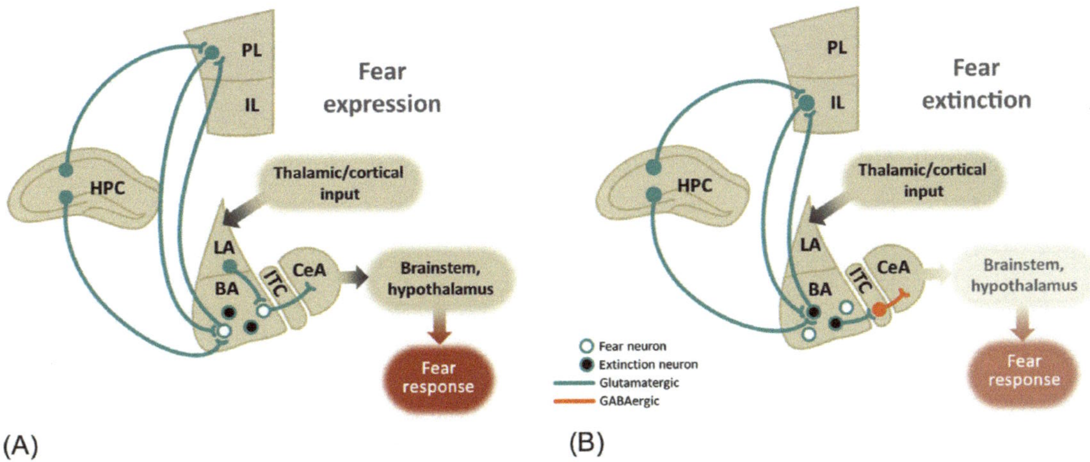

Fig. 2 Neural circuitry of fear expression and extinction. (A) During fear expression thalamic and cortical inputs conveying information about the CS and the US arrive first at the LA, which excites the CeA through stimulation of a particular neuronal subpopulation in the BA known as "fear neurons," eliciting fear responses via successive projections to brainstem and hypothalamic sites. The BA receives also excitatory input from PL cortex, thereby promoting the expression of conditioned fear. (B) During fear extinction the IL cortex excites the GABAergic ITC neurons, either directly or through activation of the "extinction neurons" subpopulation in the BA. ITC neurons inhibit the CeA, thereby inhibiting conditioned fear and promoting extinction. In addition, the HPC modulates both PL and IL as well as amygdalar activity, integrating contextual information processing. *BA*, basal amygdala; *CeA*, central amygdala; *HPC*, hippocampus; *IL*, infralimbic prefrontal cortex; *ITC*, intercalated cells; *LA*, lateral amygdala; *PL*, prelimbic prefrontal cortex. *(Modified with permission from Flores et al. (2015).)*

represents its main output region responsible for the expression of fear responses. It projects to downstream structures in the brainstem and the hypothalamus, thus launching the endocrine, autonomic, and emotional fear responses indicated previously. The formation of fear memories also requires the participation of the prelimbic division of the PFC (dorsal anterior cingulate in humans), which has excitatory projections to the BA and constitutes a key pathway for the cortical modulation of fear (Sierra-Mercado, Padilla-Coreano, & Quirk, 2011).

The extinction of fear recruits similar brain regions to those involved in fear expression, but instead of the prelimbic PFC, fear extinction engages the infralimbic subdivision of the PFC (ventromedial PFC in humans) (Sierra-Mercado et al., 2011). Indeed, prelimbic and infralimbic subdivisions of the PFC have opposing influences on fear processes, since the infralimbic cortex is required for fear extinction but not fear acquisition, and prelimbic stimulation is able to impair fear extinction (Sierra-Mercado et al., 2011). Interestingly, the BA is involved in both the formation and the extinction of fear memory, and two distinct populations of projecting neurons have been identified in this brain structure: (1) the so-called fear neurons, which are active during fear conditioning and show reciprocal connections with the prelimbic PFC; and (2) the "extinction neurons," which respond during fear extinction and are connected bidirectionally with the infralimbic cortex (Herry et al., 2008; Senn et al., 2014). Thus infralimbic PFC neurons project to extinction neurons within the BA, which activate a series of intraamygdala circuits (involving a specific GABAergic population of neurons called intercalated cells) that converge in the inhibition of the central amygdala and suppress the expression of fear. The hippocampus has also robust reciprocal connections with the BA (Pitkänen, Pikkarainen, Nurminen, & Ylinen, 2000) and is essential for fear extinction, since the new association CS-noUS that is formed during extinction is linked to the context where extinction training takes place (Myers & Davis, 2007). It has been proposed that the hippocampus activates the infralimbic cortex and certain inhibitory neurons in the BA during exposure of the animal to the extinction context, thus inhibiting output neurons in the central amygdala and preventing fear expression (Quirk & Mueller, 2008).

PHARMACOLOGICAL MODULATION OF FEAR EXTINCTION

As previously described, considerable effort has been recently made in identifying neural circuits underlying the formation and extinction of fear.

Moreover, numerous studies have investigated the neurochemical and molecular substrates of fear memories. These investigations have provided important new information for the development of a variety of pharmacological tools regulating fear processing. Next, we summarize some of the main pharmacological interventions targeting neurotransmitter systems involved in extinction facilitation.

Glutamatergic signaling plays a crucial role in synaptic plasticity and many forms of learning and memory, including fear extinction. Several studies have examined the effects of the partial NMDA receptor agonist D-cycloserine (DCS) on extinction in both rodents and humans. In preclinical models, most of the studies on DCS suggest facilitation of extinction, whether it is administered systematically or within local amygdalar or hippocampal circuits (Fitzgerald, Seemann, & Maren, 2014). DCS represents one of the best examples of translational research. Thus DCS augmentation of the cognitive behavioral therapy has been clinically studied in several anxiety disorders including posttraumatic stress disorder, phobias, or panic disorder (Singewald, Schmuckermair, Whittle, Holmes, & Ressler, 2015). Besides glutamate, the inhibitory neurotransmitter GABA also plays a role in extinction learning. Modulation of GABAergic signaling in the basolateral amygdala can enhance fear extinction. However, infusion of both $GABA_A$ antagonists and agonists in this brain area (Fitzgerald et al., 2014) has been shown to enhance extinction depending on the precise timing of the drug injection in relation to extinction training.

Three important monoamine neurotransmitter systems (serotonin, noradrenaline, and dopamine) are also involved in the regulation of extinction. Selective serotonin reuptake inhibitor (SSRI) antidepressants can modify extinction learning. Thus chronic treatment with fluoxetine, but not citalopram, enhances extinction in rodents (Burghardt & Bauer, 2013). Moreover, clinical studies performed so far have demonstrated that the combination of SSRI with behavioral therapy could be beneficial for panic disorder patients and possibly for posttraumatic stress disorder and social anxiety (Singewald et al., 2015). The noradrenergic system is crucial for both the formation and the maintenance of fear memories, as well as for fear extinction. Noradrenaline can enhance neuronal excitability in key brain regions for extinction such as the infralimbic PFC (Mueller, Porter, & Quirk, 2008), and successful fear extinction is associated with enhanced extracellular levels of noradrenaline in the medial PFC (Hugues, Garcia, & Léna, 2007). In agreement, many studies show that systemic or local (infralimbic PFC or amygdala) administration of the nonselective β antagonist propranolol impairs extinction learning (Fitzgerald et al., 2014). An increasing amount of evidence suggests that dopamine signaling has an important role in extinction mechanisms. The dopaminergic system innervates forebrain extinction circuits through mesocorticolimbic projections from the ventral tegmental area. The administration of the dopamine precursor L-Dopa enhances fear extinction in mice and healthy humans. Congruent with this, L-Dopa increases the neuronal activity in the infralimbic PFC of mice and the functional coupling of the medial PFC in humans (Singewald et al., 2015).

The endocannabinoid system may also play an important role in the modulation of fear extinction (Marsicano et al., 2002). Most of the studies suggest that enhancement of endocannabinoid signaling through activation of the CB_1 receptor facilitates extinction learning whereas CB_1 blockade with drugs such as the antagonist rimonabant impairs such learning (Fitzgerald et al., 2014).

Various neuropeptides and their corresponding receptors are located in brain regions mediating emotional behavior and have been involved in the modulation of fear extinction. For example, neuropeptide S, a neurotransmitter of ascending brainstem cells, can facilitate fear extinction when administered into the amygdala in rodents (Jüngling et al., 2008). Neuropeptide Y, an abundant peptide in the CNS, promotes extinction. In agreement, an impairment of fear extinction has been observed in mice lacking neuropeptide Y (Verma, Tasan, Herzog, & Sperk, 2012).

Other targets such as neurotrophins BDNF and FGF2, glucocorticoids, L-type-calcium channels, and epigenetic modifications have been also shown to modulate fear extinction processing (Singewald et al., 2015). Moreover, recent investigation has revealed that the hypothalamic neuropeptide orexins play an important role in the regulation of emotional memory, as described in the next section.

MODULATION OF FEAR BY OREXINS

A growing body of evidence demonstrates that the orexin system is involved in the regulation of fear processing, consistent with the existence of bidirectional projections with the limbic system. Clinical studies have revealed that patients with narcolepsy, a pathology associated with a loss of orexin neurons (Peyron et al., 2000), showed no amygdala activation and no increase in functional coupling between the amygdala and the medial PFC in response to an aversive conditioned stimulus (Ponz et al., 2010). A microdialysis study in humans has also shown an increase of orexin-A levels in the amygdala during positive emotions, social interaction, and anger, behaviors that induce cataplexy in narcoleptic patients (Blouin et al., 2013).

Studies in animal models have clearly confirmed that orexins are involved in the regulation of emotional memory. Mice lacking the OX1R showed impaired freezing behavior and reduced expression of *zif268* (an immediate-early gene which is considered a marker of neuronal activation) in the LA in both cued and contextual fear conditioning paradigms (Soya et al., 2013). In agreement, the OX1R antagonist SB334867 attenuated freezing levels related to contextual fear (Wang, Li, & Kirouac, 2017), although a recent study did not observe changes in fear learning in orexin-deficient mice (Khalil & Fendt, 2017). Moreover, the dual orexin antagonist almorexant reduced fear-potentiated startle responses (Steiner, Lecourt, & Jenck, 2012), and increased prepro-orexin mRNA levels were positively correlated with immobility time after footshock exposure in rats (Chen, Li, & Kirouac, 2014). Orexin neurons project to the locus coeruleus and locus coeruleus noradrenergic neurons strongly express OX1R. Notably, restoration of OX1R expression in noradrenergic neurons of the locus coeruleus by using an adeno-associated virus vector normalized cued fear behavior and increased LA activity after cued testing in OX1R knockout mice (Soya et al., 2013). A similar involvement of OX1R in the locus coeruleus in amygdala-dependent threat learning was revealed by pharmacological and optogenetic approaches (Sears et al., 2013). In this study, pharmacological blockade of the OX1R in the locus coeruleus by microinjection of the OX1R antagonist SB334867 impaired threat memory formation induced by an auditory stimulus. Moreover, optical stimulation of orexin fibers in the locus coeruleus was sufficient to enhance this response (Sears et al., 2013). These data are congruent with the previously established implication of the locus coeruleus noradrenergic neurons in emotional memory formation. Moreover, these results suggest that the circuit involving orexin-locus coeruleus-LA pathway might be involved in the formation of emotional memory (Soya et al., 2017).

In contrast to the results about OXR1 discussed previously, mice lacking OX2R showed reduced freezing behavior in a contextual fear test, but normal freezing in a cued fear paradigm (Soya et al., 2013). Consistent with this, intracerebroventricular infusion of the OX2R antagonist TCS-OX2-29 did not modify freezing responses in a cue-induced fear conditioning paradigm (Sears et al., 2013). These data suggest a different role for OX1R and OX2R in the formation of cued and contextual fear memories probably due to the different location of these receptors in brain structures such as the amygdala and hippocampus.

Orexins also modulate the extinction of acquired aversive memory. OX1R blockade with SB334867 facilitated the consolidation of fear extinction in both contextual and cued tests, while orexin-A infusion impaired this response (Flores et al., 2014). Direct injection of SB334867 within the basolateral amygdala mimicked the enhancement of fear extinction previously observed after systemic administration of this compound. The OX1R antagonist also rescued contextual and cued fear extinction in the 129S1/SvImJ inbred mouse strain (Flores et al., 2014), a well-established mouse model of impaired extinction. In agreement with these data, the activation level of hypothalamic orexin neurons appears to be negatively correlated with successful extinction of cue-conditioned fear in rodents (Sharko, Fadel, Kaigler, & Wilson, 2017). These data suggest the potential usefulness of this pharmacological target in diseases characterized by inappropriate maintenance of aversive memories. On the contrary, OX2R blockade with the antagonist TCS-OX2-29 or orexin-B infusion did not modify fear extinction (Flores et al., 2014).

Several studies measuring immediate-early gene expression conclude that hypoactivity of specific brain areas of the cortico-amygdala circuit, such as the infralimbic PFC and the basolateral amygdala, is related to impaired fear extinction (Holmes & Singewald, 2013). On the contrary, complete extinction of conditioned fear is associated with increased immediate-early gene expression in these brain areas (Herry & Mons, 2004). Notably, the extinction-facilitating effects of SB334867 were related to increased activity of both infralimbic PFC and basolateral amygdala, suggesting that OX1R blockade facilitates the activation of pro-extinction circuits in specific cortical and amygdalar regions (Flores et al., 2014). In agreement with this idea, a neuronal circuit associated with the facilitation of extinction induced by the OX1R antagonist SB334867 has been recently identified. Thus this antagonist increased the recruitment of basolateral amygdala projection neurons targeting the infralimbic PFC during fear extinction trial (Flores, Herry, Maldonado, & Berrendero, 2017), consistent with previous observations suggesting that "fear neurons" and "extinction neurons" overlap with two different anatomically defined subtypes of basolateral amygdala neurons projecting to prelimbic PFC or infralimbic PFC, respectively (Senn et al., 2014) (Fig. 3). Moreover, SB334867 increased the activation of GABAergic interneurons expressing the calcium-binding protein calbindin within the basolateral amygdala (Fig. 3). In agreement, activation of

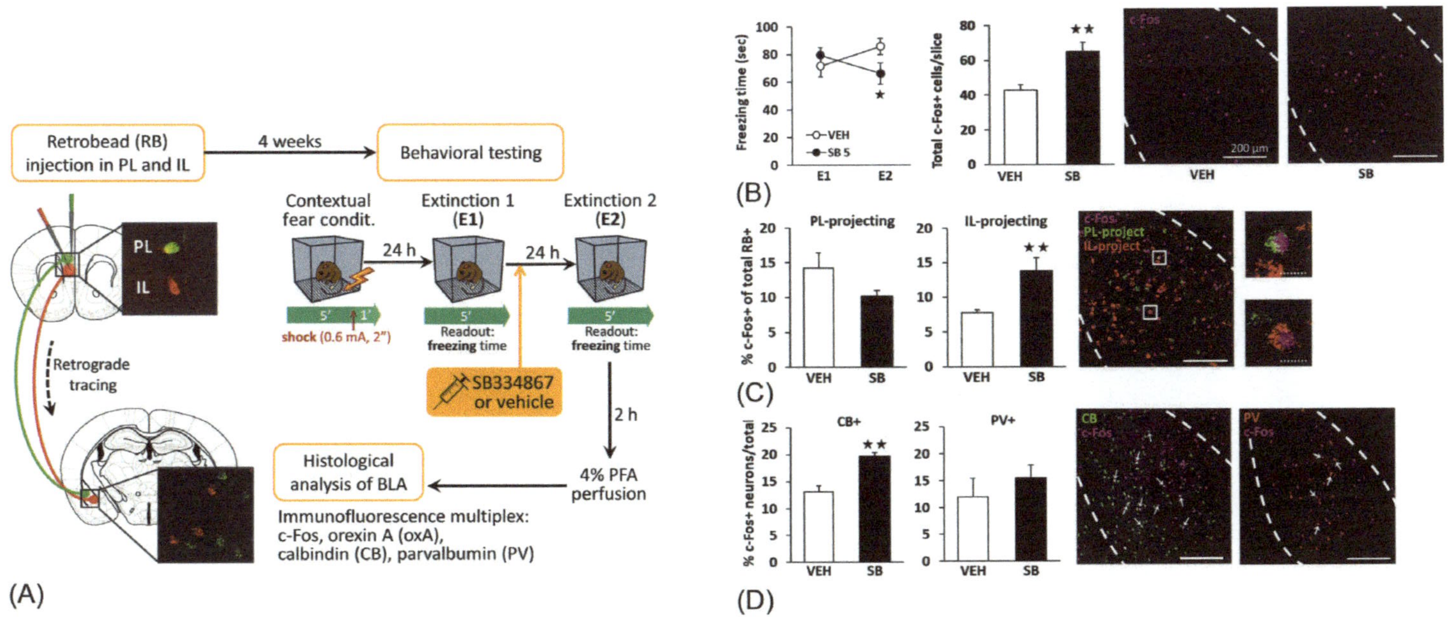

Fig. 3 OX1R antagonism facilitates fear extinction and activates specific amygdala cell subpopulations. (A) Experimental design. Animals were injected with differently colored retrobeads into the prelimbic (PL) and infralimbic (IL) prefrontal cortex. After 4 weeks mice underwent contextual fear conditioning and freezing behavior was scored 24h later during contextual reexposure (extinction training 1, E1). Immediately after E1, mice were treated with the OX1R antagonist SB334867 (SB) or vehicle (VEH), and freezing behavior was measured again 24h later (extinction session 2, E2). Animals were processed for histochemical analysis 2h after E2. (B) *Left panel*: time spent freezing by VEH- and SB-treated mice during E1 and E2. *Right panel*: Number of c-Fos+ neurons within the BLA in VEH- and SB-treated mice after the E2. (C) Percentage of PL- and IL-projecting neurons expressing c-Fos in mice receiving VEH or SB. Representative images of c-Fos *(purple)* and retrobead *(red* and *green)* colocalization at BLA are also shown. (D) Percentage of calbindin+ (CB) and parvalbumin+ (PV) interneurons expressing c-Fos in mice receiving VEH or SB. Representative images showing c-Fos *(purple)* expression within CB+ *(green)* and PV+ *(red)* neurons are also displayed. *BLA*, basolateral amygdala. *Arrows* indicate coexpression of c-Fos. *(Modified with permission from Flores et al. (2017).)*

calbindin-expressing interneurons appears to be inversely correlated with fear expression (Butler et al., 2011). Thus rats exposed to predator odor displayed increased fear responses and reduced c-Fos expression in calbindin-positive neurons in the basolateral amygdala (Butler et al., 2011). At present, whether this increased activity is related to the modulation of the basolateral amygdala-infralimbic PFC circuit remains to be elucidated.

As previously mentioned, recent evidences indicate that the PVT, a subregion of the dorsal midline thalamus, is involved in conditioned fear processing. This thalamic nucleus strongly innervates both prelimbic and infralimbic portions of the PFC, as well as the basal and central nuclei of the amygdala. Temporary inactivation of midline thalamic nuclei by microinjecting the GABA agonist muscimol (Padilla-Coreano, Do-Monte, & Quirk, 2012) and bilateral lesions of the thalamic paraventricular nucleus (Li, Dong, et al., 2014; Li, Hu, and de Lecea, 2014) after conditioning decreased freezing behavior produced by a tone-conditioned fear. However, a recent study showed no effect on freezing behavior in both cued and contextual fear conditioning paradigms by the injection of a dual orexin receptor antagonist within the PVT (Dong, Li, & Kirouac, 2015). This result suggests that orexin receptors in the PVT are not involved in the expression of conditioned fear.

In summary, orexins are critically involved in the neural mechanisms mediating emotional memory formation during fearful situations. Once acquired, orexin neurons are engaged to preserve fear during the normal physiological process of fear extinction through a mechanism that directly involves the activation of OX1R. Indeed, OX1R antagonism accelerates the extinction of aversive memories.

CONCLUSIONS

Although an important part of fear responses are innately programmed and well conserved across species, the process of learning is crucial to successfully overcome new potential dangers and be able to identify predictors of danger. Given that associative fear memory lies at the root of fear and anxiety disorders, understanding the neurobiological mechanisms that mediate long-term storage, retrieval, and weakening of fear memories is of great interest. Orexins have emerged as a key regulator of aversive memory, and several recent reports support their role in this process at diverse levels. Thus OX1R signaling promotes the expression of learned fear by enhancing both the acquisition and consolidation phases of fear memory formation, a process in which orexins may modulate noradrenergic activity from the locus coeruleus. On the other hand, orexins also hinder the ability to extinguish these fear memories through OX1R transmission. In this case, orexins might be regulating the communication between the basolateral amygdala and the medial PFC, impeding the infralimbic-induced facilitation of fear extinction. However, our knowledge about orexinergic modulation of emotional behavior is scarce and further research is needed. Importantly, human studies have reported that orexin dysfunction causes deficits in aversive learning, and increased activity of the orexin system is associated with acute anxiety states. These findings underline the translational value of the described preclinical results and highlight the potential of OX1R antagonists as promising opportunities for treating anxiety disorders characterized by abnormal fear persistence, such as posttraumatic stress disorder and phobias.

REFERENCES

Blouin, A. M., Fried, I., Wilson, C. L., Staba, R. J., Behnke, E. J., Lam, H. A., et al. (2013). Human hypocretin and melanin-concentrating hormone levels are linked to emotion and social interaction. *Nature Communications*, *4*, 1547.

Burghardt, N. S., & Bauer, E. P. (2013). Acute and chronic effects of selective serotonin reuptake inhibitor treatment on fear conditioning: implications for underlying fear circuits. *Neuroscience*, *247*, 253–272.

Butler, R. K., Sharko, A. C., Oliver, E. M., Brito-Vargas, P., Kaigler, K. F., Fadel, J. R., et al. (2011). Activation of phenotypically-distinct neuronal subpopulations of the rat amygdala following exposure to predator odor. *Neuroscience*, *175*, 133–144.

Chen, Q., de Lecea, L., Hu, Z., & Gao, D. (2015). The hypocretin/orexin system: an increasingly important role in neuropsychiatry. *Medicinal Research Reviews*, *35*, 152–197.

Chen, X., Li, S., & Kirouac, G. J. (2014). Blocking of corticotrophin releasing factor receptor-1 during footshock attenuates context fear but not the upregulation of prepro-orexin mRNA in rats. *Pharmacology, Biochemistry, and Behavior*, *120*, 1–6.

Dong, X., Li, Y., & Kirouac, G. J. (2015). Blocking of orexin receptors in the paraventricular nucleus of the thalamus has no effect on the expression of conditioned fear in rats. *Frontiers in Behavioral Neuroscience*, *9*, 161.

Fitzgerald, P. J., Seemann, J. R., & Maren, S. (2014). Can fear extinction be enhanced? A review of pharmacological and behavioral findings. *Brain Research Bulletin*, *105*, 46–60.

Flores, Á., Herry, C., Maldonado, R., & Berrendero, F. (2017). Facilitation of contextual fear extinction by orexin-1 receptor antagonism is associated with the activation of specific amygdala cell subpopulations. *The International Journal of Neuropsychopharmacology*, *20*, 654–659.

Flores, Á., Saravia, R., Maldonado, R., & Berrendero, F. (2015). Orexins and fear: implications for the treatment of anxiety disorders. *Trends in Neurosciences*, *38*, 550–559.

Flores, Á., Valls-Comamala, V., Costa, G., Saravia, R., Maldonado, R., & Berrendero, F. (2014). The hypocretin/orexin system mediates the extinction of fear memories. *Neuropsychopharmacology*, *39*, 2732–2741.

Ganella, D. E., & Kim, J. H. (2014). Developmental rodent models of fear and anxiety: from neurobiology to pharmacology. *British Journal of Pharmacology*, *171*, 4556–4574.

Herry, C., Ciocchi, S., Senn, V., Demmou, L., Müller, C., & Lüthi, A. (2008). Switching on and off fear by distinct neuronal circuits. *Nature*, *454*, 600–606.
Herry, C., & Johansen, J. P. (2014). Encoding of fear learning and memory in distributed neuronal circuits. *Nature Neuroscience*, *17*, 1644–1654.
Herry, C., & Mons, N. (2004). Resistance to extinction is associated with impaired immediate early gene induction in medial prefrontal cortex and amygdala. *The European Journal of Neuroscience*, *20*, 781–790.
Hofmann, S. G. (2008). Cognitive processes during fear acquisition and extinction in animals and humans: implications for exposure therapy of anxiety disorders. *Clinical Psychology Review*, *28*, 199–210.
Holmes, A., & Singewald, N. (2013). Individual differences in recovery from traumatic fear. *Trends in Neurosciences*, *36*, 23–31.
Hugues, S., Garcia, R., & Léna, I. (2007). Time course of extracellular catecholamine and glutamate levels in the rat medial prefrontal cortex during and after extinction of conditioned fear. *Synapse*, *61*, 933–937.
Jüngling, K., Seidenbecher, T., Sosulina, L., Lesting, J., Sangha, S., Clark, S. D., et al. (2008). Neuropeptide S-mediated control of fear expression and extinction: role of intercalated GABAergic neurons in the amygdala. *Neuron*, *59*, 298–310.
Khalil, R., & Fendt, M. (2017). Increased anxiety but normal fear and safety learning in orexin-deficient mice. *Behavioural Brain Research*, *320*, 210–218.
Kindt, M. (2014). A behavioural neuroscience perspective on the aetiology and treatment of anxiety disorders. *Behaviour Research and Therapy*, *62*, 24–36.
Kukkonen, J. P., & Leonard, C. S. (2014). Orexin/hypocretin receptor signalling cascades. *British Journal of Pharmacology*, *171*, 314–331.
Lang, P. J., Davis, M., & Ohman, A. (2000). Fear and anxiety: animal models and human cognitive psychophysiology. *Journal of Affective Disorders*, *61*, 137–159.
de Lecea, L., Kilduff, T. S., Peyron, C., Gao, X., Foye, P. E., Danielson, P. E., et al. (1998). The hypocretins: hypothalamus-specific peptides with neuroexcitatory activity. *Proceedings of the National Academy of Sciences of the United States of America*, *95*, 322–327.
LeDoux, J. E. (2000). Emotion circuits in the brain. *Annual Review of Neuroscience*, *23*, 155–184.
LeDoux, J. E. (2007). The amygdala. *Current Biology*, *17*, R868–R874.
Li, Y., Dong, X., Li, S., & Kirouac, G. J. (2014). Lesions of the posterior paraventricular nucleus of the thalamus attenuate fear expression. *Frontiers in Behavioral Neuroscience*, *8*, 94.
Li, J., Hu, Z., & de Lecea, L. (2014). The hypocretins/orexins: integrators of multiple physiological functions. *British Journal of Pharmacology*, *171*, 332–350.
Mahler, S. V., Moorman, D. E., Smith, R. J., James, M. H., & Aston-Jones, G. (2014). Motivational activation: a unifying hypothesis of orexin/hypocretin function. *Nature Neuroscience*, *17*, 1298–1303.
Marcus, J. N., Aschkenasi, C. J., Lee, C. E., Chemelli, R. M., Saper, C. B., Yanagisawa, M., et al. (2001). Differential expression of orexin receptors 1 and 2 in the rat brain. *The Journal of Comparative Neurology*, *435*, 6–25.
Marsicano, G., Wotjak, C. T., Azad, S. C., Bisogno, T., Rammes, G., Cascio, M. G., et al. (2002). The endogenous cannabinoid system controls extinction of aversive memories. *Nature*, *418*, 530–534.
Mueller, D., Porter, J. T., & Quirk, G. J. (2008). Noradrenergic signaling in infralimbic cortex increases cell excitability and strengthens memory for fear extinction. *The Journal of Neuroscience*, *28*, 369–375.
Myers, K. M., & Davis, M. (2007). Mechanisms of fear extinction. *Molecular Psychiatry*, *12*, 120–150.
Padilla-Coreano, N., Do-Monte, F. H., & Quirk, G. J. (2012). A time-dependent role of midline thalamic nuclei in the retrieval of fear memory. *Neuropharmacology*, *62*, 457–463.

Peyron, C., Faraco, J., Rogers, W., Ripley, B., Overeem, S., Charnay, Y., et al. (2000). A mutation in a case of early onset narcolepsy and a generalized absence of hypocretin peptides in human narcoleptic brains. *Nature Medicine*, *6*(9), 991–997.

Peyron, C., Tighe, D. K., van den Pol, A. N., de Lecea, L., Heller, H. C., Sutcliffe, J. G., et al. (1998). Neurons containing hypocretin (orexin) project to multiple neuronal systems. *The Journal of Neuroscience*, *18*, 9996–10015.

Pitkänen, A., Pikkarainen, M., Nurminen, N., & Ylinen, A. (2000). Reciprocal connections between the amygdala and the hippocampal formation, perirhinal cortex, and postrhinal cortex in rat. A review. *Annals of the New York Academy of Sciences*, *911*, 369–391.

Ponz, A., Khatami, R., Poryazova, R., Werth, E., Boesiger, P., Schwartz, S., et al. (2010). Reduced amygdala activity during aversive conditioning in human narcolepsy. *Annals of Neurology*, *67*, 394–398.

Quirk, G. J., & Mueller, D. (2008). Neural mechanisms of extinction learning and retrieval. *Neuropsychopharmacology*, *33*, 56–72.

Riebe, C. J., Pamplona, F. A., Pamplona, F., Kamprath, K., & Wotjak, C. T. (2012). Fear relief-toward a new conceptual frame work and what endocannabinoids gotta do with it. *Neuroscience*, *204*, 159–185.

Sakurai, T., Amemiya, A., Ishii, M., Matsuzaki, I., Chemelli, R. M., Tanaka, H., et al. (1998). Orexins and orexin receptors: a family of hypothalamic neuropeptides and G protein-coupled receptors that regulate feeding behavior. *Cell*, *92*, 573–585.

Sears, R. M., Fink, A. E., Wigestrand, M. B., Farb, C. R., de Lecea, L., & Ledoux, J. E. (2013). Orexin/hypocretin system modulates amygdala-dependent threat learning through the locus coeruleus. *Proceedings of the National Academy of Sciences of the United States of America*, *110*, 20260–20265.

Senn, V., Wolff, S. B., Herry, C., Grenier, F., Ehrlich, I., Gründemann, J., et al. (2014). Long-range connectivity defines behavioral specificity of amygdala neurons. *Neuron*, *81*, 428–437.

Sharko, A. C., Fadel, J. R., Kaigler, K. F., & Wilson, M. A. (2017). Activation of orexin/hypocretin neurons is associated with individual differences in cued fear extinction. *Physiology & Behavior*, *178*, 93–102.

Sierra-Mercado, D., Padilla-Coreano, N., & Quirk, G. J. (2011). Dissociable roles of prelimbic and infralimbic cortices, ventral hippocampus, and basolateral amygdala in the expression and extinction of conditioned fear. *Neuropsychopharmacology*, *36*, 529–538.

Singewald, N., Schmuckermair, C., Whittle, N., Holmes, A., & Ressler, K. J. (2015). Pharmacology of cognitive enhancers for exposure-based therapy of fear, anxiety and trauma-related disorders. *Pharmacology & Therapeutics*, *149*, 150–190.

Soya, S., Shoji, H., Hasegawa, E., Hondo, M., Miyakawa, T., Yanagisawa, M., et al. (2013). Orexin receptor-1 in the locus coeruleus plays an important role in cue-dependent fear memory consolidation. *The Journal of Neuroscience*, *33*, 14549–14557.

Soya, S., Takahashi, T. M., McHugh, T. J., Maejima, T., Herlitze, S., Abe, M., et al. (2017). Orexin modulates behavioral fear expression through the locus coeruleus. *Nature Communications*, *8*, 1606.

Steiner, M. A., Lecourt, H., & Jenck, F. (2012). The brain orexin system and almorexant in fear-conditioned startle reactions in the rat. *Psychopharmacology*, *223*, 465–475.

VanElzakker, M. B., Dahlgren, M. K., Davis, F. C., Dubois, S., & Shin, L. M. (2014). From Pavlov to PTSD: the extinction of conditioned fear in rodents, humans, and anxiety disorders. *Neurobiology of Learning and Memory*, *113*, 3–18.

Verma, D., Tasan, R. O., Herzog, H., & Sperk, G. (2012). NPY controls fear conditioning and fear extinction by combined action on Y_1 and Y_2 receptors. *British Journal of Pharmacology*, *166*, 1461–1473.

Wang, H., Li, S., & Kirouac, G. J. (2017). Role of the orexin (hypocretin) system in contextual fear conditioning in rats. *Behavioural Brain Research*, *316*, 47–53.

CHAPTER 9

Conclusion

Joshua A. Burk*, Jim R. Fadel†
*Department of Psychological Sciences, College of William and Mary, Williamsburg, VA, United States
†Department of Pharmacology, Physiology & Neuroscience, University of South Carolina School of Medicine, Columbia, SC, United States

It is a daunting task to attempt to summarize in a few chapters the status of a field of research that has existed for over 20 years and has generated thousands of original research papers. Nonetheless, we have attempted to identify some of the highlights and controversies surrounding the role of the orexin/hypocretin system in a variety of normal physiological functions and pathological conditions. Clearly, the term "physiological integrator," originally coined by de Lecea and colleagues, is an apt description of this neuropeptide system which sits at the intersection of peripheral signals related to homeostasis and central mediators of cognition, autonomic function, and motivated behavior. As we look ahead to the next two decades of investigation in this area we anticipate a clearer understanding of the numerous physiological functions mediated by this system and the further development of clinical applications of orexin/hypocretin pharmacology. There are numerous questions that remain to be answered. What role might the orexin/hypocretin system play in normal aging or age-related cognitive decline? What is the cause of orexin/hypocretin deficiency in narcolepsy? What role might the orexin/hypocretin system play as a novel therapeutic target for conditions—from narcolepsy to psychostimulant addiction—in which the orexin/system has been implicated? What is the source and functions of orexin/hypocretin in the periphery? How does orexin/hypocretin signaling interact with colocalized neurotransmitters? Progress made toward answering these and other questions will surely lead to important new discoveries even new questions are generated. The continued development of contemporary neuroscience tools targeted toward the orexin system will be essential to answering these questions. Additionally, the development of novel pharmacological compounds, such as drugs that act at allosteric sites on orexin receptors, will be important for developing new pharmacotherapies for conditions associated with aberrant orexin processing.

The Orexin/Hypocretin System
https://doi.org/10.1016/B978-0-12-813751-2.00009-7

Finally, we wish to thank each of the chapter contributors for their investment of time in this project. We further thank the numerous students, research staff, and collaborators from our own labs and institutions for their contributions of enthusiasm, hard work, and insight as we have pursued our own research in this area. We also thank the representatives of Elsevier who approached us with this idea following the success of orexin/hypocretin symposium at the 2016 Society for Neuroscience meeting in San Diego. It has been a pleasure to work with Sam Young, Joslyn Chaiprasert-Paguio, and April Farr. Thank you for your commitment to this project, your patience, and for keeping us on task. Finally, we thank our families for their constant support and encouragement.

INDEX

Note: Page numbers followed by *f* indicate figures, and *t* indicate tables.

D

E

F

G

H

FSC
www.fsc.org
MIX
Papier | Fördert
gute Waldnutzung
FSC® C083411

Kolibri 360 GmbH
Ferdinand-Jühlke-Straße 7
99095 Erfurt, Deutschland
produktsicherheit@kolibri360.de